AF470503

COMMANDING OFFICERS

COMMANDING OFFICERS

David Miller

JOHN MURRAY
Albemarle Street, London

Contents

Illustrations

The author and publishers would like to thank the following for permission to reproduce illustrations: Plate 1, V&A Picture Library, London; 2, William Gilkerson, courtesy of the United States Naval War College Museum; 3, USAF; 4 and 22, US Department of Defense; 5, Mrs Margaret Pawlyn; 6, RAF Museum, Hendon; 7, 9 and 10, RN Submarine Museum, Gosport; 8, Imperial War Museum; 11 and 12, Fleet Air Arm Museum; 13, The Devonshire and Dorset Regiment; 14, Stato Maggiore Marine, Roma; 15 and 16, National War Museum of Malta via Joseph Caruana, Esq; 17, P&O Archive; 19, Trustees of the National Maritime Museum, London; 20, US Naval Institute; 21, U-Boat Archiv; 23, Colonel Bob Stewart; 24, Ton Class Association; 25, Yarrow. Plate 18 is from the author's collection.

DIAGRAMS

Preface

This book examines the role of the most important person in the chain of command in all armed forces: the commanding officer. It does so by illustrating the various facets of the subject through historical examples that show how commanding officers have coped with particular situations. These examples are tailored to illustrate the point being made; some are short and straightforward, others are, of necessity, longer and presented in greater depth to explain the complexities of the task facing an individual and the significance of his actions and reactions. Maps and tables are included, where necessary, to clarify events and to give outline details of weapons or ships to enable the reader to make comparisons of size or capability.

There are thousands of examples of commanding officers upon which to draw, and in the course of research several hundred were examined in detail. Of those, about thirty are presented here. Inevitably some readers will feel that a number have been included that would better have been omitted, while others will consider that more deserving examples have been overlooked. The requirements placed upon a commanding officer and the talents needed to meet them are, however, universal. Thus, the examples have only been drawn upon as representative, and the choice of particular individuals must not be interpreted to suggest that any one branch of the armed forces or any one nation has a greater or lesser proportion of good or bad commanding officers than any other. Clearly there is insufficient space to satisfy all claims for inclusion, but the selection has been entirely mine and I take full responsibility for it.

A number of conventions are used throughout the book, which are explained here to obviate the need for repetition.

All times use the 24-hour system and are given in local time, unless specifically stated otherwise.

In lists, items such as people, countries and ships are given in alphabetical order according to the initial letter of their names in English.

All units and measurements are given in the form used at the time. However, since in this context absolute precision is unnecessary, units are rounded to the nearest significant number. Thus, a precise measurement of 45 feet 3 inches is shown as 45 feet.

It is difficult to find a suitable collective term that embraces all the people in all units, including officers, non-commissioned officers and seamen, soldiers and airmen, and covering both men and women; and it would be tedious for the reader if these were constantly spelt out. So, where it is necessary to use a collective noun, and in the absence of any more suitable term, I have used the word 'troops'.

It used to be the case that military forces were either at peace or at war, the latter being a clearly defined, legal state of hostilities with a foreign power. Today, however, there are many situations in which units can be engaged in military operations within a situation well short of war, covering a range of activities from 'aid to the civil power' (itself ranging from the quelling of riots to disaster relief), through counter-insurgency and anti-terrorist operations, to peace-keeping and peace enforcement. Again, there is no convenient collective term, so I have used the word 'combat'. Thus, for example, 'training for combat' means preparation for any type of military activity from counter-terrorism to general war.

A book such as this has involved advice and assistance from a number of people, and I wish especially to thank the following: John Ashby, for his advice and for permission to draw on his book *Seek Glory, Now Keep Glory* concerning the 'two colonels' at Le Cateau in 1914; Brian Crabbe, for his advice on the sinking of the *Khedive Ismail*; my brothers, Squadron-Leader Robin Miller for his advice on airforce matters and for help in locating the photograph of 103 Squadron, Christopher for obtaining a map which had long eluded me, and both of them for proofreading some of the chapters; Phillip Haythornthwaite for drawing my attention to Lieutenant-Colonels Beven, Elkington and Mainwaring; Rear-Admiral Horace B. Robertson, Jr., formerly of the United States Navy's Judge-Advocate General's Corps for his advice on the USS *Nicholas* affair and for obtaining for me a copy of the edited version of the naval investigative report; Colonel Bob Stewart, DSO, for his advice and comments on the manuscript concerning his adventures in Bosnia in 1992–3 and for providing a picture of himself at work in Bosnia; Dr John Sweetman of the Royal Military Academy, Sandhurst, for permission to draw on his book *The Dambusters' Raid* for facts concerning Wing-Commander Guy Gibson and 617 Squadron, RAF; and Mrs Shirley Westrup, former Chairman of the 103 Squadron Association, for advice on matters relating to that squadron in which she served during the Second World War.

In the years in which I have been researching this subject I have discussed matters with a variety of experts, friends and colleagues, some of whom have been kind enough to comment on various entries. However, the views and opinions expressed in the book are mine alone.

1

A Many-talented Officer

Among the plethora of books about great – and not so great – navy, army and air leaders at various levels of command, there is one remarkable gap: there are very few books about commanding officers. This is all the more surprising because they are the most important links in the entire chain of command. These are the officers who train units in peace, who lead them in war and who are, in a unique and very personal way, totally responsible for their success or failure.

The basic, permanently constituted and self-contained fighting unit in all armed forces is the naval warship, the army battalion and the airforce squadron. Such units are normally part of larger formations, such as fleets and flotillas, divisions and brigades, or air groups and wings, but it is the warship, the battalion and the squadron that fight the battles.

There is, of course, a vast range of commanding officers' appointments in all forces: navies have shore units, armies have artillery, engineer, communications and logistical units, and airforces have other types of flying units, such as transport squadrons, as well as ground units specializing in tasks such as engineering, communications and maintenance. All these units pose their own challenges, and require particular attributes and professional skills in their commanding officers. This book, however, is concerned with the basic fighting unit; that is to say, the naval warship, whether surface ship or submarine, the army and marine infantry battalion, and the airforce fighter and bomber squadron.

The commanding officer is carefully selected and appointed to his specific unit by higher authority, and he is given a number of well-defined responsibilities and duties that are set out in a multitude of legal requirements, administrative rules and service regulations. These range from how to maintain discipline to the supervision of accounting, but they also include a host of practical instructions – the United States Navy, for example, being one of many to advise that 'if it becomes necessary to abandon the ship, the Commanding Officer should be the last person to leave'. Such rules and regulations are, however, only the starting point. The commanding officer must also possess a number of far more important, albeit often intangible, qualities.

There is no one blueprint for command, and the exercise of it is a matter that varies not only from one individual to another, but also from one service to another, even from one nation to another. A brief survey of some of the better-known senior leaders in the Second World War will serve to make the point.

Two of the most famous army commanders in the United States Army were George Patton, who was brash, extrovert, thrusting and a born risk-taker, and Omar Bradley, a quiet, unassuming and altogether more cautious man. Among the Supreme Allied Commanders, Douglas MacArthur was aristocratic, arrogant and aloof, while Dwight Eisenhower was homely, modest and more at home as a coalition leader than as a field commander. All four were great and successful commanders, but they were quite different in character and employed quite different methods of command and leadership; and all four in their turn differed from the Navy's Chester Nimitz, who was bold and imaginative, but certainly not flamboyant in the MacArthur or Patton style.

In the German Wehrmacht, Gerd von Runstedt was cold, aloof, deep-thinking and a careful tactician, while Erwin Rommel was bold, thrusting and much more down to earth. By contrast the Navy's Karl Dönitz was an intensely ambitious and calculating man who was, nevertheless, extremely popular with his U-boat men. Then, too, in the British Army, Bernard Montgomery was brash and self-important but essentially cautious; quite different from Bill Slim, a bluff, good-natured and transparently honest man, who led the campaign in Burma.

Each of these men was supremely competent, but each was unique not only in character and personality but also in the way in which he applied his military skills and exercised command. Thus, if there are no stereotypes for admirals or generals, there is unlikely to be one for commanding officers. Nonetheless, there are certain attributes, traits and skills which can contribute, in varying combinations, to a successful commanding officer, and there are lessons which can be learnt from the experiences of others.

The environments in which commanding officers serve vary considerably. A warship is a self-contained community which, whether surface ship or submarine, includes all the seamen, engineers, communicators, maintenance men and others necessary for it to go to war. The commanding officer in such a warship is traditionally a person of great, almost complete, authority, being described in the Royal Navy, for example, as the 'first after God'. Ships traditionally spent a long time at sea, interspersed with brief and usually very bloody periods of combat, and the commanding officer demanded absolute obedience in the latter and steady discipline in the former. A failure in either respect had to be dealt with firmly and speedily; except in rare and extremely serious cases, there was no question of keeping a man in chains for months until the ship next reached port.

Surface warships vary in size from a corvette with a crew of, perhaps, 200, to an aircraft-carrier, with about 6,000 aboard, and the rank of the commanding officer varies from lieutenant to captain accordingly. The responsibilities are, however, the same, the only difference being one of scale: the size of the ship, the number of men and women in the crew, and the complexity of the equipment on board.

The commanding officer of a submarine is, if anything, even more powerful, having sole authority over every aspect of the submarine's handling and over the lives of the crew while on patrol. Traditionally, the only man with unrestricted access to the periscope, he is the sole decision-maker in all tactical matters and his crew survive or die by his skill. Such responsibilities are demanding enough in modern submarines, which can be at sea for up to seventy days at a time, all bar a few hours of which are spent submerged and with only the most rudimentary communications with the outside world. But these modern boats are large and comparatively spacious compared with those used in the Second World War. Then a German crew could spend up to a year aboard a very much smaller U-boat on a voyage to and from the Far East.

In land warfare the principal combat unit in both the army and the marines is the infantry battalion, which in virtually all nations is commanded by an officer with the rank of lieutenant-colonel. Such battalions sometimes operate on their own, but more often they are part of a brigade or other force. Even so, the commanding officer is still autonomous. The basic battalion includes its own engineers, communicators and maintenance elements, albeit on a small scale, but if it is operating independently it will have been given additional elements to make it self-supporting.

With only a few exceptions (for example, helicopters), airforces operate from a fixed airfield shared by two or more squadrons. The airfield is commanded by a base commander, usually a colonel or group captain, while the squadrons are commanded by lieutenant-colonels or wing-commanders.

While many of the requirements and problems affecting these commanding officers are similar, there are also some substantial differences. The naval commanding officer is in the ship or submarine with the rest of the ship's company and they all, quite literally, sink or swim together. The army commanding officer is in a generally similar position in that he is with his troops and will, almost always, share the same fate as them. The airforce commanding officer, on the other hand, has to divide his time between the base, where there are many important matters demanding his attention, and air operations in which he leads the combat element of his unit into battle.

Despite these different environments and their varying requirements, there are a number of attributes that are required of all commanding officers. A full list would be very long, but the factors that follow are considered to be the most important, although, as always, each individual will possess them to differing degrees.

Perhaps the most important quality of all, leadership, is by its very nature elusive. One Royal Air Force officer, for example, in describing his commanding officer, Wing-Commander Gibson, as 'the outstanding influence on the squadron', added that 'though we did not see a lot of him he seemed to set a standard of perfection in all our training and the final preparation. It's called leadership – how do you define it?'[1] How, indeed? Some elements can be learnt in a classroom and others acquired by experience, while the armed forces assist in bolstering authority by conferring rank and imposing regulations, under which disobedience incurs punishment. But true leadership, which is particularly important in a commanding officer, is more than all this and includes elements of personality, character and example that depend upon the individual.

One particular requirement of military leadership lies in the ability to motivate subordinates to do things which, viewed rationally and dispassionately, they might well not wish to do. Such tasks vary from the merely arduous or unpleasant, through the dangerous to the probably fatal, the last being aptly described by the eighteenth-century British admiral Lord Howe when he remarked that 'some occasions in our profession will justify, if not require, more hazard to be ventured than can be systematically defended by experience'. The American and British bomber crews operating from the United Kingdom in 1942–4 would have known what he meant. They continued to fly even though they knew full well that the probability of surviving a full tour (thirty missions for the British, twenty for the Americans) was considerably less than 50 per cent. Nor could they close their eyes to the losses, since the evidence was all around them. In combat they saw their comrades shot down. Back at base, small groups of airmen would go round the aircrew's accommodation, collecting the belongings of those who had failed to return.

The reasons which keep each individual going in such circumstances are varied. Sometimes it is simply a matter of patriotism, the traditional cry of 'King and Country' or the 'Old Flag'. Some fight for an ideal. Lieutenant-Commander Eugene Esmonde, the British officer who died attacking German battlecruisers in the English Channel in 1942, was an intensely religious man who saw himself as part of a holy crusade, fighting against the forces of evil. Others are equally motivated, but they fight less for moral reasons than for the 'honour' of the ship, regiment or squadron. Then there are those who fight simply because they have to and would be punished if they did not. Whatever the underlying motives, however, it is the commanding officer who has to pull all the elements of his unit together and ensure that they fight as one coherent whole.

A commanding officer must possess the tactical and organizational skills necessary to handle the unit under his command and the technical knowledge to deploy its assets with maximum effectiveness. In part this is a question of mechanics. A warship commanding officer needs to be a first-class ship handler, an army officer must be capable of managing all the elements of his battalion in both space and time, while the airman must be a good pilot. But the best sort of commanding officer is capable of much more than this, for he possesses that certain touch that enables him to handle his unit boldly, to take the enemy by surprise and to achieve results that are out of proportion to the size of the force involved. Thus, the commanding officer will be the master of events rather than their servant, finding radical solutions that sometimes break what others regard as tactical 'rules', but which enable him to impose his will on the battlefield.

Like leadership, courage is hard to define, but there are two elements which would be included in any list. The most obvious is physical courage – the ability to overcome fear and to accept danger. Troops will soon notice if the commanding officer is lacking in such courage, though this does not mean that he should not know fear; indeed, someone who is genuinely 'fearless' is a rather dangerous person to be near. Instead, the courageous person is one who recognizes and understands fear but has the strength to overcome it. Except in a few instances, troops do not generally require their commanding officer to be foolhardy, and they may become resentful if they feel that unnecessary risks are being taken with their lives.

The second element is moral courage, the inner strength that enables a commanding officer to accept responsibility and to drive through something which may not at first meet with the approval of his troops. In several of the examples that follow, commanding officers implemented challenging training programmes as soon as their units were temporarily withdrawn from combat, not to deprive them of rest but in order to maintain the standards and practise the skills that they knew were necessary. A further element is the strength to insist on doing what is morally right, when the easier course is to allow something else to happen – in other words to establish the limits of what is acceptable and then to maintain them.

There is no doubt that the responsibilities of command impose a substantial burden on an individual. He therefore needs to possess a combination of mental and physical robustness that will enable him to endure the strain of battle. Where commanding officers are concerned, youth is somewhat less important than ability and experience. One of the most successful and respected commanding officers of the 2nd Battalion of the Devonshire Regiment during the First World War was 25 years old when he was killed in action in 1916, while Wing-Commander Guy Gibson, who was specially selected to form and lead 617 Squadron in 1943, was 24 years old at the time of the Dambusters' raid. Of greater significance than their youth, however, was not only their exceptional ability, but also the fact that they had been serving since 1914 and 1939 respectively and therefore had as much experience as considerably older men of the new form of conflict in which they were involved.

At the other end of the age scale there can be no doubt that, while there will clearly be variations between individuals, there comes a time when a person is no longer fit for the hectic and arduous life of a commanding officer. This poses a particular problem at the start of wars, since the age of commanding officers tends to creep up during times of peace: the average age of commanding officers in the British Army in 1914, for example, was the late forties. The result is that, in the early days of a conflict, some men find themselves in situations where they are unable to cope with the mental and physical demands which have suddenly been imposed upon them.

There is another side to the question of robustness in that a commanding officer must be able to make hard decisions when he knows that he is right – for example, to persist with a plan when it appears to be ill-advised or going wrong, to give up a plan when he considers that the effort is going to be futile, or even, in some extreme circumstances, to sacrifice the lives of a few for the benefit of the many.

The attributes that are part of the personal make-up of a commanding officer will have been assessed during his career and will have played a key part in his selection for the post. Once he has become a commanding officer, there are then a number of essential factors concerning the unit over which he has varying degrees of control.

A fundamental responsibility of the commanding officer is the morale of his unit, since from that everything else will flow. How this is achieved and what is necessary to sustain it will vary according to the commanding officer's personality, the individual backgrounds of the people in the unit, the service and the nation. Thus, for example, if a selection were to be made of infantry battalions of the British, French, German and United States armies whose morale was high, it would be found that the way in which their commanding officers had achieved this and the weight they had given to the various factors that contribute to morale might be quite different. There are even differences within nations. For example, in the French Army different measures are needed to ensure high morale within a battalion of the Foreign Legion and a line battalion, or in the British Army within the Parachute Regiment, a battalion of the Guards and a county regiment of the line.

One important factor in morale is that of confidence. Individuals need to have confidence in themselves, in their unit and, in particular, in the commanding officer. A second necessary element in morale is success, since the men will want to belong to an organization that is successful – and success in one field quickly leads to success in others.

Although outside the commanding officer's control, the public's perception of their country's armed forces is also important to the morale of those troops. Thus, for example, overwhelming public support for the British forces in the Falklands War was a great morale booster, while American units in the latter days of the Vietnam War knew that there was considerable opposition to what they were doing among certain sections of the public at home, knowledge that resulted in a significant lowering of morale among those units in the field.

The overall organization of the unit and its men by rank and specialization will be decreed by higher authority and it is very unlikely that the commanding officer will have much say in the matter except, possibly, in the case of a few individuals. That being so, the commanding officer's main functions are to trim the organization to meet his requirements, to deploy the officers and troops as best he may to take the greatest advantage of their talents, and to get rid of the few that do not fit in.

Disciplinary rules are codified in legal documents that set out what constitutes a military offence, how it is to be tried and how it is to be punished. However, it will be up to the commanding officer to set the tone within his unit and to 'shade' what is and is not acceptable. Thus what is right in disciplinary terms for one unit may well be quite inappropriate in another.

Although the final decision will always be his, it is important that the commanding officer listens to subordinates, especially those who are expert in fields with which he himself is less familiar. The skill lies in accepting advice where it is sound and rejecting it where it is not. One particular duty with regard to his subordinates is that the commanding officer must select and train the next generation of leaders, a factor which adds considerably to his power, since the subordinates know they depend largely upon him for the reports and recommendations that will ensure their promotion.

Under modern conditions there can be no escape from the media, for whom the commanding officer is the natural focal point. Some commanding officers clearly wish for nothing more than that the media should go away, but it is an unavoidable fact of life in modern operations and commanding officers have a stark and simple choice: the media is either their friend or their enemy.

Lastly, a large number of intangible factors affects the commanding officer, of which four in particular are significant.

Command is frequently a lonely task, the very appointment setting the commanding officer apart from his fellows and bestowing on him, whether at sea, on land or in the air, special status. As in other respects, there are differences between branches of the armed forces. In armies and airforces commanding officers live among other officers, whereas in navies they are set apart, nowhere more so than in the Royal Navy where the captain leads a totally separate existence, with his own accommodation and even eating alone on most occasions. Whatever the physical arrangements, however, the responsibilities of the position set the commanding officer apart.

There are differences in the way in which commanding officers can be held to account – or hold themselves to account – for failure. Indeed, some commanding officers in the nineteenth and twentieth centuries took their responsibilities so seriously that, when they felt they had brought disgrace or disaster upon their unit, they took their own lives. This applied particularly at sea where some captains shot themselves, while others followed the rather unusual sea-going practice of 'going down with the ship'.

Many commanding officers who have achieved fame, and deservedly so, have done so for a glorious event which lasted a few hours or, at most, a few days. Of equal merit, but far less famous, are those who have kept their units going through the long hard slog of a campaign. It required a very particular set of capabilities, for example, to take a battalion forward time after time for a spell in the trenches during the First World War or to lead a battalion of the United States Marines in the Pacific in the Second World War in yet another landing against utterly determined Japanese, who would have to be killed to the last man before the island was firmly conquered.

There are also occasions when commanding officers are placed in a position where their unit is obliged to fight to the 'last man and last round'. Sometimes this is as a result of a deliberate order, but more often it is because the unit has become trapped. In both cases there is usually an option for individuals of the unit to surrender, but frequently they do not do so.

Perhaps the greatest challenge of all is that frequently a commanding officer will have little idea of what is going to happen next. In an infantry battalion in a conventional campaign, for example, one attack may be succeeded by another attack, or a defence, or a move to another part of the battlefield. The challenges are even greater in so-called peacetime where a commanding officer can move from a conventional war exercise to a counter-terrorist operation and thence to peacekeeping in a United Nations force. Some such changes may be the result of months of advance planning, while others arise, almost literally, overnight. As will be described, on one quiet day in 1975, Lieutenant-Colonel Austin's Marine Corps battalion was undergoing training on Okinawa. Forty-eight hours later he and his troops were on an island a thousand miles away, which none of them had heard of before, and in the midst of a violent operation in which sixteen marines were killed and forty-three wounded.

In another case, although he received rather more advance warning, Lieutenant-Colonel Bob Stewart had to move his battalion from the relatively familiar environment of the North German Plain and preparation to repel an armoured attack from eastern Europe, to a United Nations operation in Bosnia in 1993 where both the setting and the mission were completely unfamiliar. Similarly, naval and air forces find themselves facing new and unexpected challenges. A warship may move in a matter of hours from attacking submarines to shelling a land target and then defending itself against attack by a surface vessel; or, in peace, from taking part in a conventional naval war (as in the Gulf) to supporting peacekeeping operations ashore or rescuing people involved in natural disasters, such as earthquakes or floods.

2

'A Genius for War'

Perhaps the most basic requirement of a commanding officer is that he should be good at his job, because those serving under him will soon discover if he is not, and their confidence will wax or wane accordingly. Commanders who possess this felicitous talent have an instinctive feel for the tactical handling of their unit, coupled with the ability to command the respect and willing obedience of their troops. Such people bring a sure and certain touch to the conduct of their operations. The Germans describe this as *Fingerspitzengefühl*, (literally, 'finger-tip feeling'). Colloquially it means flair or a sixth sense that enables them, almost instinctively, to seize on the right thing to do, or to be at the right place at the right time, either to ensure victory or to stave off defeat.

Such a quality was possessed in abundance (though only at a tactical level) by the Second World War German general, Erwin Rommel. The English language does not possess a directly equivalent expression, but Napier, one of the most distinguished writers on the Peninsular War, probably came closest to it when he described one of the Duke of Wellington's battalion commanders as 'having a genius for war'.

Lieutenant-Colonel John Colborne

The Battle of Waterloo, 18 June 1815

An Englishman, John Colborne was born in 1778 and commissioned into the 20th Regiment of Foot in 1796. He campaigned in the Netherlands, Minorca, Egypt and Malta before being appointed military secretary (that is, personal staff officer) to Major-General Sir John Moore in 1806. Colborne was present at Moore's death at the Battle of Corunna in 1808, following which he returned to England, where he was promoted to major. He went back to the Peninsula under Wellington later that year and in December 1809 was promoted to lieutenant-colonel, achieving his ambition of commanding a battalion, in this case the 2nd Battalion, 66th Regiment.

Wellington quickly came to have great faith in Colborne's abilities, as a result of which his battalion was frequently given the most difficult tasks and positions. Then, in July 1811, Colborne transferred to take command of the 52nd Regiment of Foot, one of the light infantry regiments that had been trained by Sir John Moore. One of his first exploits with his new battalion was at the siege of Ciudad Rodrigo, in which he took just twenty minutes to capture the outlying fort of San Francisco. An onlooker attributed his success to

> the clear conception and explanation of the plan of attack, so that each individual in charge knew what he had to do; the high discipline and order in which the plan was carried out under the eye of the officer commanding the party [Colborne]; and the care taken to cover the redoubt with a sheet of fire.

Such tactical precepts, here being employed in the first decade of the nineteenth century, remain valid today. Indeed, the divisional commander, Major-General Craufurd, a notoriously taciturn and hard-to-please man known as 'Black Bob', was heard to mumble that 'Colonel Colborne seems to be a steady officer'; no man serving under him ever received higher praise.

Colborne returned to England again for a short period to recover from a wound but was soon back in the Peninsula. There, in 1813, when riding out with another famous soldier, Captain Harry Smith, he encountered a column of 400 French troops. Colborne dispatched Smith to bring up the battalion, which was some distance away, but, realizing that it was vital to seize the initiative, he then rode up to the French commander and firmly demanded his immediate surrender. Assuming that the confident Englishman must have troops in close support, the astonished Frenchman agreed.

Following the end of the Peninsular campaign, Colborne and the 52nd moved to the Netherlands and were thus already on the ground when Napoleon escaped from Elba and then marched on Brussels. At Waterloo the 52nd Regiment, fighting fit, well trained, well armed and with implicit trust in its commanding officer, was part of Adam's Brigade which, for the first few hours of the battle, was held in reserve on the Allied right wing, near the hamlet of Braine l'Alleud. The battle started at about 1100 on 18 June 1815, but it was not until about 1500 that the brigade was ordered forward to a position above the farm of Hougoumont, which had already been the scene of some of the stiffest fighting of the day. There the 52nd found itself in a huge cornfield that had been reduced to a boggy morass as a result of three days' heavy rain and the manoeuvring of thousands of men and horses since early morning. To make matters worse, the ground was littered with the wounded and dead of both sides, with injured and dead horses, abandoned and damaged guns and equipment, and the general detritus of battle. In addition to this, there was a deafening cacophony of sound: shouted orders, wounded men screaming, muskets and artillery firing, drums rolling and bugles blaring.

Colborne formed the battalion, about a thousand men strong, into two squares and then awaited developments. At about 1730 Wellington ordered Colborne to retire up the hill out of range of some French artillery, but Colborne declined, telling the staff officer who brought the message that the artillery was not troubling the battalion. A little later, however, Colborne saw men of the Nassau Regiment fleeing the woods to his front and decided that the time had come to withdraw.

As the battalion retired, Colborne followed on horseback, between the rear rank and the enemy. At that moment, a colonel of cuirassiers deserted the French line and galloped up to Colborne, showing that he was changing sides by shouting *'Vive le Roi!'* The Frenchman pointed to Napoleon, telling Colborne that the Imperial Guard was on the march and was about to make a grand attack. Like every other soldier in Europe, Colborne knew that the Emperor only deployed his famous *corps d'élite* at the critical moment in battle. Almost simultaneously, the Imperial Guard emerged from the French position in columns of fours, a formation they had used for fifteen years, since it combined an element of shock with the close camaraderie and mutual encouragement needed by conscript armies. However, although the formation had impressed many Continental armies, the phlegmatic British infantry had never been especially awed by the threat of the Imperial Guard, nor were they on this occasion.

On reaching the crest of the ridge, Colborne halted his battalion, turned it about to face the enemy and aligned it in two ranks to increase its firepower, taking it upon himself to disregard Wellington's order that it should be four deep. By now it was clear that the column of the Imperial Guard, some 6,000–7,000 strong, intended to attack the battalion immediately to the 52nd's left – the British 1st Regiment of Foot Guards. Seeing the long column of the Imperial Guard, with every soldier's eyes intent on the target to his front, Colborne immediately saw an opportunity and, entirely on his own initiative, gave the order for the 52nd to advance, and to wheel to the left as it did so.

The 52nd was in two ranks and, with a strength of 1,000 men less some 200 men engaged in skirmishing, this gave a frontage of approximately 400 men. Colborne now gave the orders

1. The 52nd Regiment of Foot at Waterloo, 18 June 1815

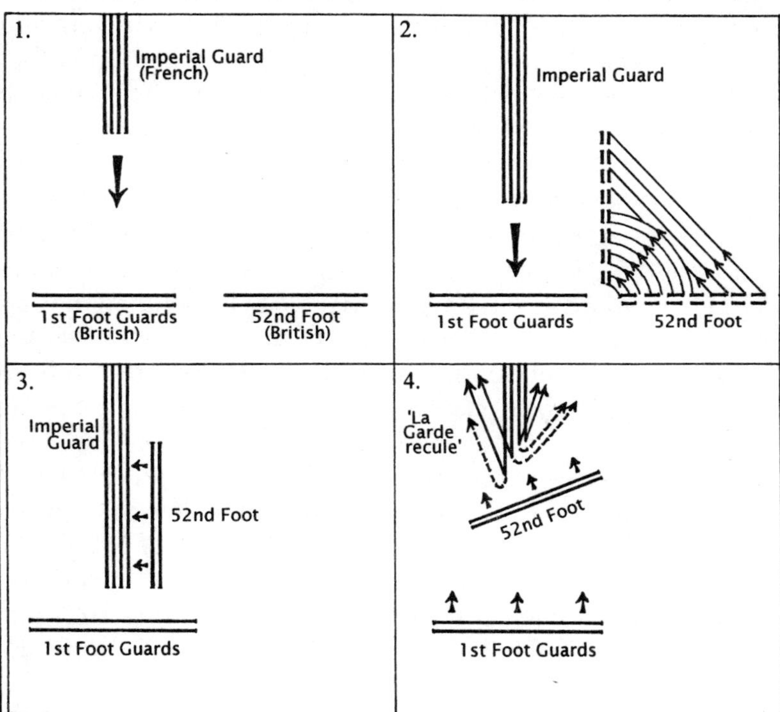

1. The 1st Regiment of Foot Guards (British) are the right-hand battalion of Byng's Brigade and are starting to run out of ammunition. The 52nd Foot is immediately to the right of the 1st Foot Guards in two ranks. The Imperial Guard (French) approaches in columns of companies, each four ranks deep.

2. Colborne orders the 52nd to wheel to the left. The two left-hand companies wheel, pivoting on the left-hand man. The two right-hand companies 'left form', that is they turn half left and then advance straight towards their position in the new formation.

3. Having paused briefly to straighten the line, Colborne orders the 'charge'. The 52nd moves along the front rank of the 1st Foot Guards, hitting the French Imperial Guard along its left flank.

4. The Imperial Guard breaks up, turns and flees. The 52nd then changes direction to the right and follows in pursuit.

for the two left-hand companies to 'wheel' to their left in line, while the right-hand companies 'formed' to their left. This entailed the two left-hand companies changing direction through 90°, while maintaining their formation, which meant that the outside men had to move fast to keep up. The other two companies also changed direction through 90°, but with each man heading direct for his new position and each moving at the same speed.

Battalions normally used either one manoeuvre or the other, but Colborne, very unusually, combined the two, a daring and complicated tactic that would have daunted almost any battalion on the parade ground in peacetime, never mind on the battlefield. In addition, Colborne took an immense risk with the right flank of his battalion, which was wide open to fire from the main French position and which did, in fact, suffer quite heavy casualties.

The task was not, however, impossible. The 52nd continued with its turn until it was facing the flank of the French Imperial Guard at a distance probably slightly in excess of 100 yards. Colborne paused briefly to straighten the line and then ordered the battalion to advance towards the enemy, with skirmishers deployed in front. This caused the 52nd to move along the front of the (British) 1st Guards on their left. Then, judging his moment to have come, Colborne gave the order to charge. The 52nd ran forward with a will, even though their progress was temporarily impeded by having to cross a sunken lane. Being charged was a new and unpleasant experience for the Imperial Guard and after a brief pause it was soon in full retreat. Within minutes, the rest of the French army, some crying incredulously, '*La Garde recule!*', was also in retreat. Colborne again wheeled the battalion, this time to the right, and advanced towards the main French position, maintaining the momentum past Napoleon's recently deserted headquarters at La Belle Alliance until they met their Prussian allies just as darkness fell. Then, exhausted by their efforts, the 52nd halted for the night.

Colborne was exceptionally confident in his own abilities,

but this was not an arrogant and aggressive self-confidence, rather a quiet and modest self-possession. Thus, he felt quite able not only to decline Wellington's instruction to withdraw up the hill, but also to seize the tactical initiative without delay. Had he waited to refer his proposal to higher authority, the opportunity would have gone for ever. For if the Imperial Guard had broken through the British line, Napoleon would almost certainly have reached Brussels the following day.

Colborne's success was not solely due to his being battle-wise. He also looked after his men, which they realized and appreciated. He ensured that they carried twice as much ammunition and powder as laid down in regulations and thus they never ran short of either in battle. He also ensured that his officers took a proper interest in their men, that food of as good quality as circumstances permitted was supplied regularly, and that the best possible arrangements were made for the wounded.

Over six feet tall and always turned out smartly, Colborne was 'a soldier's soldier'. His courage was beyond question and he took the most extraordinary personal risks, suffering numerous wounds in the process. Despite this, he was also a thoughtful man, who gave the most careful consideration to all his actions. As a result, at the time of the Peninsular and Waterloo campaigns, when Wellington had forged the units under his command into one of the most efficient and close-knit forces Britain has ever produced, Colborne was widely recognized as 'the finest battalion commander in the British Army'. Even though he later reached the rank of field marshal, was created a baron and was awarded decorations by many countries, there is probably no other epitaph by which Colborne himself would have wished to be remembered.[1]

LIEUTENANTS JOHN PAUL JONES AND NICHOLAS BIDDLE

Actions against the Royal Navy, 1776

Lieutenants John Paul Jones and Nicholas Biddle were among the first officers to be commissioned into the Continental Navy during the American War of Independence and both were masters of their craft, being not only effective and courageous fighting men but also skilful and experienced mariners. Both found themselves in situations where they were faced by a much more powerful enemy, but their solutions to their dilemma were highly original and also showed that 'luck' can play a vital role in the outcome.

John Paul Jones

Lieutenant John Paul Jones took command of the 14-gun sloop *Providence* in February 1776. *Providence* was just 110 feet long and had a crew of seventy-three officers and sailors, plus twenty-five marines, which together represented a sizable pro-portion of the total assets then available to the rebel navy. Even so, Congress was determined that it should be used aggres-sively. On 21 August 1776, Jones was given orders to 'proceed immediately on a Cruise against our Enemies and we think in and about the Latitude of Bermuda may prove the most favourable for your purpose'. His specific task was to 'seize, take, sink, burn or destroy' British shipping.

On 1 September, Jones sighted five British ships sailing to New York and headed straight for the enemy, but he was no sooner committed than a lookout spotted HMS *Solebay*, a 28-gun frigate, that was much larger and considerably more pow-erful than the American sloop.

Jones quickly concluded that there was nothing to be gained by an unnecessary confrontation with a greatly superior enemy, ordered *Providence*'s helm put down and started to

2. John Paul Jones Outsails the Captain of HMS *Solebay*, 1 September 1776

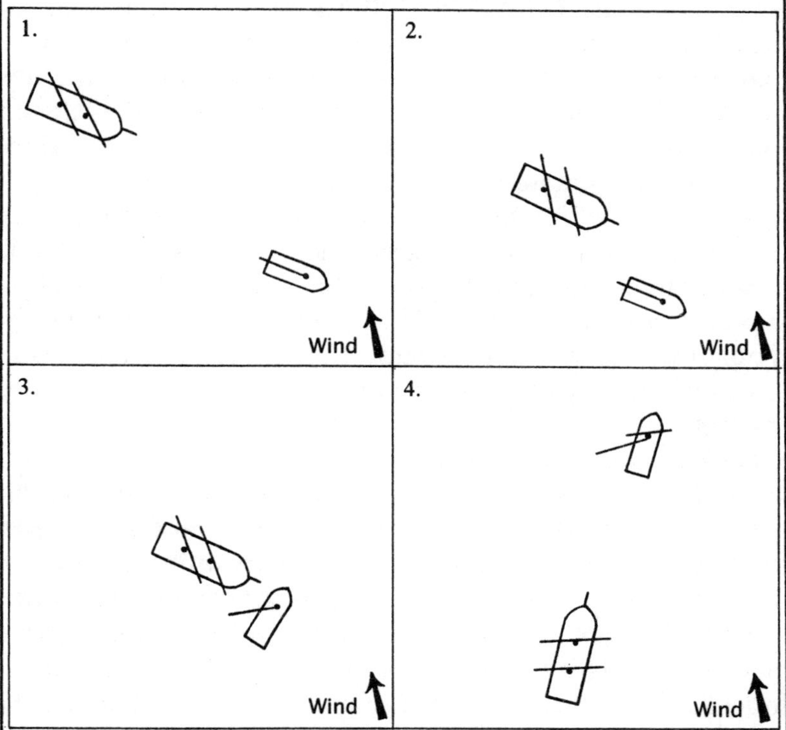

1. The pursuit starts with *Providence* fleeing close-hauled to windward, with *Solebay* some distance astern.

2. By late afternoon, *Solebay* is almost within cannonshot of the smaller and much weaker *Providence*.

3. *Solebay* is almost ready to open fire when Jones puts down his helm, lets out the main boom, breaks out every square-sail the sloop possesses, cuts under *Solebay*'s bowsprit and runs before the wind.

4. By the time *Solebay*'s captain has realized what has happened, turned into *Providence*'s wake and reset his sails, the smaller boat is well away, quickly disappearing into the gathering darkness.

escape. *Solebay*'s commander, Captain Thomas Symonds, had other ideas and proved to be particularly persistent. The chase lasted all day, with the two captains employing all their considerable skills. Despite the efforts of the Americans, the British man-of-war drew steadily closer, and by late afternoon there was no more than a musket shot between them, although the Americans still held the windward gauge, *Providence*'s fore-and-aft rig having enabled her to sail a point closer to the wind than the square-rigged frigate.

As the British frigate approached, Jones discussed the situation with his executive officer, Lieutenant Rathburn, and together they hatched a plan to take advantage of their sloop's superior manoeuvrability. Jones prepared by sending the topmen aloft to spread the spars and to ready the running sails – topsails, topgallants and studdingsails – and then awaited his moment.

For his part, Captain Symonds started with a *ruse de guerre*. When *Providence* broke her American colours, the British discharged two guns harmlessly to leeward, a traditional signal for 'I am friendly'. Jones, however, knew full well that no 'friendly' vessel would have carried out such an unremitting pursuit. With *Solebay* close under his lee and a broadside imminent, he also knew that he would not have a second chance. Indeed, if the British fired just one effective broadside, he would have no option but to strike his colours.

At that point, facing imminent defeat, he spotted some cat's-paws appearing to windward, ruffling the wavetops. Experience told him that a small squall was approaching. He waited for the precise moment and then ordered the helmsman to put the helm up, swinging *Providence* directly across *Solebay*'s bows and running so close under her bowsprit that she was 'within pistol shot'. Captain Symonds was taken completely by surprise and failed to respond in time, so that *Providence*'s sails filled and she hauled away on the new course, her speed increasing with every second as she ran before the wind, which, as Jones well knew, was the sloop's best point of sailing.

The British, for whom success had seemed so close, had been wrong-footed. Symonds was an experienced officer, but his reaction was more instinctive than reasoned. He tried to turn inside the American, which proved to be his undoing, since he was hit by the squall that Jones had spotted and, it would seem, that Symonds had not. As a result, the frigate's sails were briefly taken aback, adding to the confusion on board. The final problem for the British was that the port (leeward) battery had not been manned, and thus when *Solebay* passed under the American sloop's stern she was unable to fire a single cannon until it was too late.

This gave Jones a few precious minutes' grace. He took full advantage of the situation to gain a headstart on the new tack. By the time the British had sorted themselves out, the 'rebel' was showing them a clean pair of heels, eventually disappearing into the darkness as night fell.

Nicholas Biddle

Jones's success in the *Solebay* incident stands in stark contrast to that of his fellow American captain, Nicholas Biddle, commander of the 32-gun frigate *Randolph*. On 8 March 1778, Biddle was escorting a small convoy when he was intercepted by the British frigate HMS *Yarmouth*, which, with 64 guns, was exactly twice the size of his own vessel. *Yarmouth* sighted the convoy at 1730 and gave chase, coming within range of *Randolph* at 2100. Biddle hoisted the American colours and the *Randolph* exchanged broadsides with *Yarmouth* for some fifteen minutes, fighting very hard. Eventually, Biddle realized that such tactics must inevitably result in his ship's destruction, so he decided to attempt to take the enemy by boarding. The American was closing to implement this audacious plan when a chance shot found his magazine and his ship blew up.

The damage to *Yarmouth* subsequently recorded in her log makes it obvious that the *Randolph* had been extremely close. According to the log, the American ship had kept up his firing very smart for about a quarter of an hour and then blew up. It was fortunate for us that we was to windward of her. As it was, the ship was in a manner covered with part of hers, and a large piece of timber stuck fast to the topgallant sail . . . Carried away sails and rigging all to pieces. We had 5 men killed and 12 wounded. On this ship blowing up, the rest [that is, the other ships in the American convoy] departed in different ways and were soon out of sight.

The British captain of *Yarmouth*, Vincent, sailed away, not thinking that anyone could survive such a violent explosion, but on passing through the area again four days later he chanced upon four survivors, whom he immediately rescued and treated with great kindness. A few days later, he put them ashore at Savannah under a flag of truce.

The actions of Jones and Biddle during the American War of Independence illustrate how fine is the line between success and failure. Neither man had any illusion that he stood much chance against a stronger opponent, and it is clear that in Biddle's case, he fought such an unequal engagement in order to secure the safety of the other ships in the convoy by gaining them the time they needed to scatter and escape. In this, Biddle's action bears many similarities to that of Captain Kennedy, commanding the British armed merchant cruiser HMS *Rawalpindi*, who was also protecting a convoy when, on 23 November 1939, the German battlecruiser *Scharnhorst* appeared over the horizon. Kennedy steamed straight towards the enemy in order to buy the merchant ships the time they needed to scatter and, like Biddle's, his ship was sunk.

Biddle's fight was extremely courageous, but the unfortunate outcome was that the infant Continental Navy lost one of its best ships, some 300 men and one of its finest commanders – 'the kind of naval captain that the God of Battles makes', as John Paul Jones later described him.

John Paul Jones was undoubtedly a highly skilled mariner, having risen from apprentice to ship's captain in a short space of time, and it was this professional skill that enabled him not only to judge *Providence*'s capabilities to a nicety, but also to detect the approach of the squall that so upset Captain Symonds and slowed *Solebay*'s turn to port. However, even before the end of the encounter, Jones displayed other qualities. The chase lasted for twelve and a half hours, during which time the British frigate – patently larger and better armed – approached ever closer. Despite the increasing tension, the American captain remained calm, quietly proceeding with his plans and calculations but also ensuring that his crew maintained their individual and collective discipline.

These two incidents also illustrate the effect of that most intangible of qualities, 'luck'. Jones and Biddle, both competent seamen and excellent commanders, fought the British with equal determination. Symonds ought to have succeeded in his engagement with Jones, while Biddle might well have carried the day against Vincent had he managed to board the *Yarmouth*. But there is no doubt that, when they both needed it, luck ran with Jones and against Biddle.[2]

COLONEL DAVID LOWNDS

The Battle of Khé Sanh, 22 January – 1 April 1968

Colonel David Lownds, United States Marine Corps, was an experienced combat officer who had commanded a platoon in the Second World War, a company in the Korean War and a battalion, the 3rd Battalion 8th Marines (3/8 Marines), in Vietnam in 1963–4. In 1967 he was selected as commanding officer of the 26th Regiment, United States Marine Corps (26 Marines). He returned to Indochina at the height of the United States' campaign, arriving on 12 August. His appointment carried with it command of the area designated Khé Sanh Combat Base.

Khé Sanh was the western anchor of the US defensive line that ran east–west 6 to 12 miles south of and parallel to the southern edge of the DeMilitarized Zone (DMZ), a supposedly neutral strip separating North and South Vietnam. The combat base itself was sited on a small plateau in the north-west corner of the Republic of Vietnam, 9 miles south of the DMZ and 2 miles east of the Laotian border. The plateau was overlooked by a semicircle of hills to the north and north-west, the highest of which was Dong Tri (3,600 feet). The combat base was also 2 miles from the village of Khé Sanh which sat astride Route 9, an all-weather road running westwards from Dong Ho on the coastal plain to the western highlands.

The terrain was rugged and hilly, and in most places the dense jungle vegetation limited ground observation to about 15 feet, while the tree canopies were so intermingled that they concealed the many trails from aerial observation. An important local factor was the weather, particularly between November and mid-March when the north-east monsoon brings the wet season. This is at its worst during the period known locally as the *crachin*, which lasts from January to April and which is characterized by low cloud and dense fog, usually accompanied by light rain or drizzle, resulting in reduced ceilings and visibility, which seriously affected operations, and in temperatures as low as 45° Fahrenheit (7° centigrade).

The enemy forces were a mixture of North Vietnamese Army (NVA) regulars and guerrillas from South Vietnam known as the Vietcong. They were commanded by General Vo Nguyen Giap, who remembered only too well his victory over the French at Dien Bien Phu in May 1954, where he had shown that revolutionary forces could defeat a Western army in a set-piece battle. In the south, the United States Military Assistance Command Vietnam (MACV) was commanded by General William Westmoreland, whose strategy was to drive the Vietcong and NVA away from populated areas and then to employ superior firepower and mobility to defeat them in the countryside. He too was aware of Dien Bien Phu as an example of what could go wrong if the enemy was allowed to seize the initiative, but he was convinced that United States forces were superior in every way to those of the French in 1954. Thus,

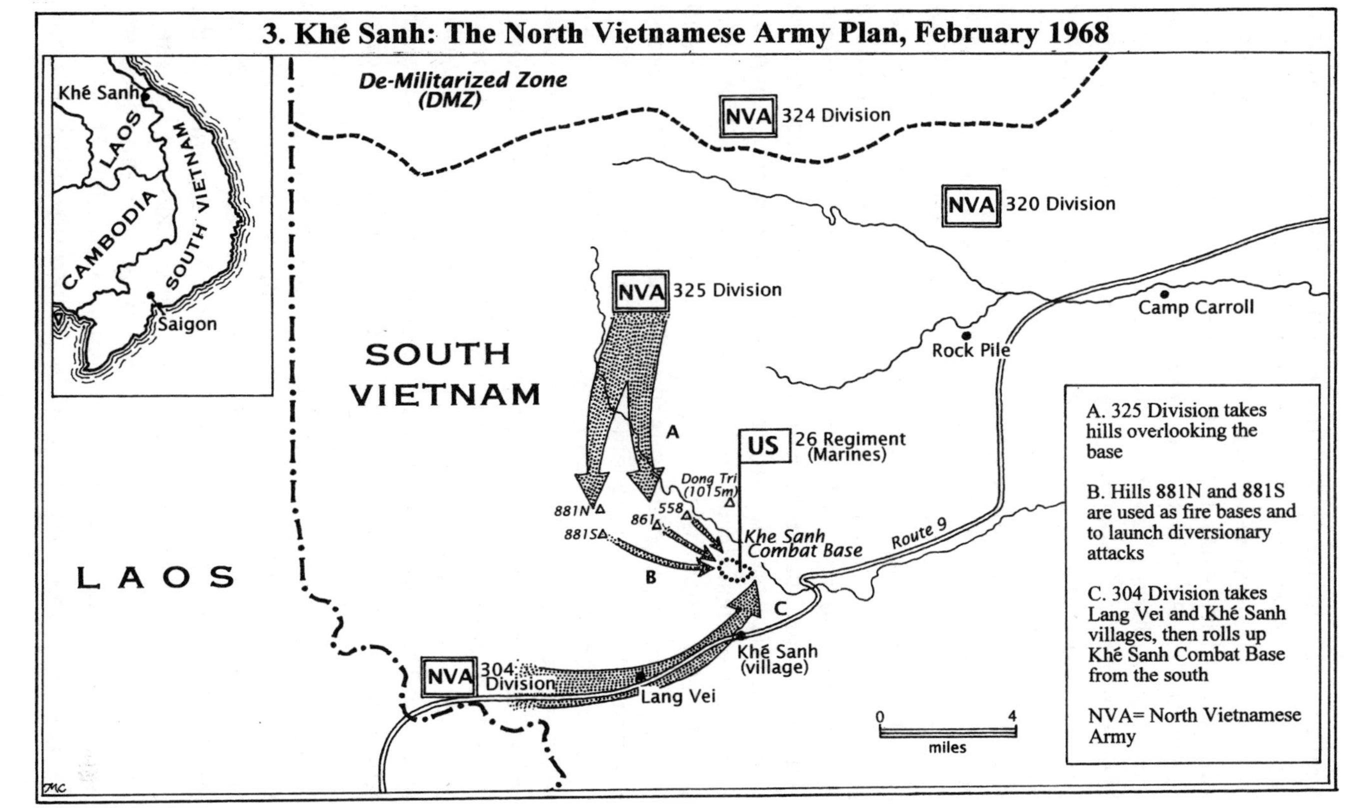

3. Khé Sanh: The North Vietnamese Army Plan, February 1968
Khé Sanh
LAOS
CAMBODIA
SOUTH VIETNAM
Saigon
De-Militarized Zone
(DMZ)
NVA 324 Division
NVA 320 Division
NVA 325 Division
SOUTH VIETNAM
Camp Carroll
Rock Pile
A
US 26 Regiment (Marines)
Dong Tri (1015m)
881N
558
861
881S
Khe Sanh Combat Base
Route 9
B
C
LAOS
NVA 304 Division
Lang Vei
Khé Sanh (village)
0
4
miles
A. 325 Division takes hills overlooking the base
B. Hills 881N and 881S are used as fire bases and to launch diversionary attacks
C. 304 Division takes Lang Vei and Khé Sanh villages, then rolls up Khé Sanh Combat Base from the south
NVA= North Vietnamese Army

when faced with the possibility of a head-on confrontation at Khé Sanh, neither commander-in-chief held back.

Khé Sanh had first been occupied in 1962, since when the strength of the garrison had fluctuated according to the perceived threat. At the time of Lownds's arrival in August 1967 it consisted of the 1st Battalion 26 Marines (1/26 Marines) and Headquarters 26 Marines. As would any incoming commander, Lownds reviewed the facilities and immediately asked that the airstrip be upgraded. This was completed by late October, giving the garrison a sound all-weather runway and taxiways that were capable of taking Lockheed C-130 Hercules transport aircraft. The equally important ground facilities for refuelling and cargo-handling were also improved. Similarly, Lownds ensured that there was good liaison with the two nearby artillery bases. The first, nicknamed 'Rock Pile', lay 6 miles to the north-east. The second and larger of the two, Camp Carroll, lay 10 miles nearer the coast. Lownds's mission was clear: to use the base to interdict the major NVA supply routes running along the Laotian side of the border and also to deter NVA incursions into the northern part of South Vietnam.

In December 1967 intelligence reported large-scale NVA troop movements towards Khé Sanh, and the garrison noticed increased activity as enemy patrols probed the base defences. In late December it was confirmed that an NVA front headquarters (equivalent to a corps headquarters in a Western army) and three enemy divisions were within striking range of Khé Sanh, with a fourth division a little further away. All had strong artillery support, and there were several armoured units equipped with Soviet PT-76 tanks, light, amphibious vehicles armed with a 76mm gun. Then, on 2 January 1968, a Marine patrol killed five enemy and found documents on the bodies that identified them as including the commander, the operations officer and the signals officer of the same NVA regiment. Clearly they had been conducting a pre-attack reconnaissance. A deserter then revealed that the NVA's 325 Division had been charged with taking Hills 881S and 861, which would be used as bases for fire support and for diversionary attacks on the northern edge of Khé Sanh while 304 Division skirted the base to the south before attacking from the east. The intention was to prevent overland reinforcements and then roll up the position. It was probably no coincidence that 304 Division had fought at Dien Bien Phu.

General Westmoreland's choice was either to hold the base, which was well sited, held by reliable and confident troops, and capable of support by massive air and artillery fire, or to withdraw the garrison, which would involve little loss of life but the abandonment of much equipment and stores. The NVA would also score a minor, if ephemeral, propaganda victory. After weighing up the factors, Westmoreland decided to hold Khé Sanh.

Lownds identified the northern hills as crucial to the defence and he reinforced them and resited the units so that every approach was dominated by fire and each position was supported by direct fire from at least one other position. This tactical sense was to pay dividends throughout the battle.

The garrison consisted of the combat forces shown in Table 1 but, in addition to these, there were communications, transport, medical and other logistical troops, together with a variety of aviation personnel from the US Marine Corps and the USAF, all of whom shared the dangers and hardships of the infantry. Altogether, the garrison at the start of the battle comprised 6,680 officers and men, who were faced by between 20,000 and 30,000 battle-hardened professionals of the NVA.

Preliminary skirmishes took place on 19–20 January but then an NVA deserter informed Lownds that the first major attack would take place on the night of 21/22 January. Forewarned, Lownds was able to repel the attack when it began just after midnight. Another, much heavier attack was launched on the following day, with an intense mortar, artillery and rocket barrage which, among other targets, hit the main ammunition dump, the fuel dump and a CS gas stockpile. The ensuing conflagration continued for forty-eight hours, with ammunition and explosives cooking off at random intervals, while gas

Commanding Officers

TABLE 1: *US Garrison at Khé Sanh, 22 January–1 April 1968*

Infantry (all US Marine Corps)
1st, 2nd and 3rd Battalions, 26th Regiment
1st Battalion, 9th Regiment
Company B, 3rd Reconnaissance Battalion
1st Platoon of Company A, 5th Reconnaissance Battalion

Infantry (Army of the Republic of Vietnam)
37th Ranger Battalion

Armour
3rd Tank Battalion:
    Company B (five M48 main battle tanks)
    Anti-tank Company (ten Ontos tank destroyers)

Artillery (at Khé Sanh)
1st Battalion (eighteen 105mm howitzers; seven 4.2-inch mortars), 13th
    Marines (Artillery)
3rd Battalion (one battery of six 155mm howitzers, towed), 12th Marines
    (Artillery)
Detachment (two M42 dual 40mm anti-aircraft guns), 44th Artillery (US
    Army)
Detachment (two M55 quad 0.5-inch machine-guns), 65th Artillery (US
    Army)
Radar detachment of 238 Counter-mortar Radar Unit (US Army)

Miscellaneous
One company of 3rd Combined Action Group (Vietnamese 'popular' forces
    with USMC advisers)
One detachment of 5th Special Forces Group (fourteen US Army 'Green
    Berets' controlling groups of Vietnamese irregulars)

Others
Other units included: engineers, communications, logistics, medical,
    aircraft maintenance, aircraft ground-handling, air traffic control,
    interrogations, postal and photographic.

Artillery (at Camp Carroll and Rockpile)
2nd Battalion of 94th Artillery (US Army) (eighteen 175mm guns)

periodically wafted over the area, making the defenders' lives unpleasant and at times hazardous. The NVA infantry attack concentrated on Hill 861 but with a subsidiary attack on Khé Sanh village, which was held by a combined action company and a South Vietnamese regional unit. Both attacks were repelled, but when the village was attacked again in the afternoon Lownds decided to evacuate the place while he could. The South Vietnamese troops were absorbed into the garrison, while the villagers were flown out to safety.

These attacks gave rise to gloomy press reports drawing parallels with Dien Bien Phu, but Westmoreland responded by reinforcing the base, with the 1/9 Marines arriving on 22 January and the 37th ARVN Ranger Battalion (37ARVN) on 27 January. A routine was quickly established: the NVA shelled the base every day and their patrols probed for weak links in the perimeter, while the Marines patrolled, filled sandbags and carried on with the never-ending task of improving the defences. The 1/9 Marines were given responsibility for the Dropping Zone (DZ) and daily sweeps cleared the edges of NVA troops, while engineers cleared the DZ itself of the inevitable crop of mines.

Meanwhile, the NVA's 1968 Tet (lunar new year) offensive had broken out, with furious battles developing in major cities such as Saigon and Hué, which were reported around the world and especially in the United States. There was a brief lull around Khé Sanh until dawn on 5 February when the NVA launched a heavy attack on Hill 861 in which they managed to penetrate the American position. They then paused, however, and a free-for-all fight developed which the Marines clearly won. A second assault on 0610 hours also ended in failure for the NVA.

The next attack on 7 February was against a nearby Special Forces camp at Lang Vei, which included large numbers of wives and children. When called upon for help, Lownds realized that an overland counter-attack would be ambushed, while the only available landing site for a heliborne attack was occupied by NVA PT-76 tanks, so he told the survivors to fight their way out. About 3,000 reached the main position at Khé Sanh, from which most were evacuated by air.

On 9 February, an NVA battalion attacked a Marine platoon position, but immediate artillery fire isolated the NVA reserves, leaving those who had reached the US positions to be dealt with by infantry counter-attacks. Then, when they withdrew, the North Vietnamese had to run the gauntlet of the US fire zones, incurring yet further casualties. This little battle cost the Marines 21 killed and 27 wounded for 150 confirmed NVA dead and 50 weapons captured.

A relative lull followed with only minor infantry actions, but the artillery duel continued. The dense morning ground mist lifted at about 1000, rising to a cloud base 100 to 500 feet above ground level and thus enabling NVA observers to direct fire on to the Americans from guns that were still concealed from observation. In the late afternoon the haze descended again and the NVA artillery quietened down. The Marines adjusted to this schedule, working in the open during the early morning and late afternoon, and taking cover during the middle of the day.

The heaviest shelling of the siege occurred on 23 February, when 1,307 incoming rounds landed in eight hours, and again an artillery dump was hit: American losses for the day were 10 killed and 51 wounded. Two days later a Marine patrol saw three NVA soldiers and gave chase, which led them into an ambush. When another platoon was dispatched to rescue the first, it too was ambushed. It took the two platoons four hours to fight their way back into the base, having lost 26 dead and 21 wounded. This was, however, one of the very few occasions when the Marines were worsted by the NVA.

On 29 February came intelligence that numerous NVA troops were heading towards Khé Sanh from the south and west, and these were targeted by all available artillery as well as by USAF B-52 bombers from Guam. These strategic bombers inflicted immense casualties, Montagnard tribesmen reporting that they had found numerous groups of up to 500 bodies of NVA troops who had been annihilated on their march into battle. Despite this, some NVA infantry attacks did take place, mostly against the 37ARVN, but these were repelled with ease.

By 1 March NVA pressure had relaxed, although artillery bombardments continued, with an average of 150 rounds daily, increasing on 23 March to 1,109. The Marines patrolled aggressively, and on 31 March the siege was declared over and the garrison was relieved, though Lownds and his Marines did not consider that they had in any way been 'rescued'. On 12 April, Colonel Lownds handed over to a new commanding officer and left Khé Sanh. His regiment was awarded a Presidential citation and he himself received the Navy Cross.

One of the features of the siege was the efficient use of the available air and artillery support. This was directed by the Fire Support Control Center (FSCC), an integrated part of Lownds's HQ, which planned and co-ordinated all fire tasks in the Khé Sanh area, whatever the means of delivery, from infantry mortars to B-52s. Under the FSCC was the Fire Direction Center (FDC) for artillery missions and the Direct Air Support Center (DASC) for air missions. Lownds kept them all on their toes, one of his practices being to walk unannounced into the FDC and designate a random spot on the map to test the gunners' responses: the average time for the first round on target was under forty seconds.

Aerial resupply was essential and, although ranges were relatively short, the bad weather at Khé Sanh, combined with hostile artillery against both airborne planes and those on the runway, made it a hazardous undertaking. Losses were incurred, and two spectacular crashes hit newspaper and TV headlines around the world, seeming to confirm the newsmen's pessimistic forecasts. Thereafter, greater use was made of low-level extraction in which aircraft flew past at a height of about 20 feet and a drogue parachute dragged the load out, which then fell to the ground. During the siege the USAF delivered 12,430 tons of supplies to Khé Sanh: 4,310 tons were air-landed and 8,120 tons were delivered by low-level extraction or normal parachute. Of this, about 3,500 tons had to be delivered to the garrisons on the northern hills by helicopter – a dangerous exercise but one that was accomplished successfully.

Between 19 January and 31 March, the Marines lost 199 men killed plus 830 who had to be evacuated because of wounds, and they withstood 70 days of complete isolation under constant fire from enemy artillery, rockets, mortars and small arms.

The North Vietnamese Army had had every intention of capturing Khé Sanh as part of their much-vaunted Tet offensive, but they grossly underestimated the determination of the garrison at Khé Sanh, the sheer power of the fire support available to the Americans, and their ability to apply it. Time and again the NVA were unable to push home an attack because the follow-up waves and reserves had been hit by air and artillery attack.

On the American side most things went right and, despite a plethora of advice and interference from his superiors, Colonel Lownds played a decisive part in the battle. The five northern hills were the tactical key to Khé Sanh and were held by the equivalent of two Marine battalions. Had the NVA taken them, it is doubtful whether the airfield position could have been held.

The NVA repeatedly laid traps for the American troops, using ambushes as bait to encourage ever greater commitment of rescue troops, a tactic which had worked very successfully both against the French in the First Indo-China War and against other American troops in the Second. At Khé Sanh, however, such tactics worked only once, when first one and then another platoon were lured into an ambush on 25 February. However, as has been noted, Lownds made the platoons fight their own way out, and the NVA ploy failed.

Lownds could not escape the knowledge that much more rested on the fate of the garrison than just its survival. Indeed, had Khé Sanh fallen and the survivors become prisoners-of-war (for North Vietnam was only a very few miles away), this would have been as severe a blow to US prestige as the fall of Dien Bien Phu had been to that of the French.[3]

'A Genius for War'

WING-COMMANDER GUY GIBSON

The Attack on the Dams, 16 May 1943

Guy Gibson was commissioned into the RAF in 1937 and served with his first bomber squadron, flying the twin-engined Handley-Page Hampden, in 1941. Having successfully completed that tour he should have gone on to a second-line appointment but he chose instead to serve with a night-fighter unit, following which he commanded another bomber unit, 106 Squadron, flying the notoriously unsatisfactory twin-engined Avro Manchester. He completed that tour on 11 March 1943, at which point he had undertaken seventy-one bomber and nine-nine night-fighter missions. He then went on leave but almost immediately received an order posting him to Headquarters 5 Group to write about his experiences as a bomber pilot. That, however, was instantly forgotten when, on arrival, he was offered a new task. In rank he was an Acting Wing-Commander and he was just 24 years old.

During the 1930s a British aircraft engineer, Dr Barnes Wallis, foreseeing the coming war, came to believe that one key to defeating Germany would lie in attacking its energy supplies among which was hydroelectric power from dams. That premise in turn drew his attention to the chief dams in the Ruhr: the Möhne, the Eder, the Sorpe and the Ennepe. His inventive brain then devised a novel method for attacking the dams in which a large mine would skip over the water until it hit the dam at a low velocity. The mine would then sink down the water-face of the dam to a preset depth, where it would explode, the water serving to concentrate the shockwaves against the structure of the dam. After much difficulty, in February 1943 Wallis eventually succeeded in getting his idea accepted by the Air Ministry. Air Chief Marshal Sir Arthur Harris, commander-in-chief of RAF Bomber Command, was ordered to form a special squadron for the operation, which would fly twenty modified four-engined Lancaster bombers. One candidate for command came instantly to mind, and on 18

March Gibson was asked whether he would volunteer for the task. Given only the haziest idea of what was involved for security reasons, he nonetheless immediately agreed. The next day he was supplied with a few more details: the unit, provisionally called Squadron X, would be based at RAF Scampton in Lincolnshire and the aircrew would be volunteers.

Gibson arrived at Scampton on 21 March. The new squadron's aircrew began to arrive on 24 March, though not all were assembled until 6 April. Some were indeed volunteers, men who knew Gibson either personally or by reputation, but a number were simply posted. Some crews came complete, while others had vacancies because one or two individuals had declined the posting. The gaps were filled on arrival at 617 Squadron. They were men of mixed experience and they were also representative of the nationalities serving in Bomber Command at that time. Of the 133 aircrew who took part in the actual operation 13 were Australian, 89 British, 28 Canadian, 2 New Zealanders and 1 from the United States. One thing they had in common, however. Until the day of the operation none of them, whether volunteers or conscripts, had the slightest idea of what was involved.

The ground crew reported between 23 and 27 March, and were simply posted in from elsewhere in 5 Group. The challenge facing them was daunting. The squadron started with ten Lancasters on loan from other squadrons, but these were, of course, the standard model and not modified to take Wallis's 'mine'. On the ground every item of equipment, from servicing trolleys to beds and chairs, had to be obtained either officially or, in many cases, unofficially, although matters were helped when the unit was officially formed on 24 March. Three days later it was given the number 617. By the end of the month virtually all the 700-odd ground crew and most of the aircrew were present, and equipment was pouring in.

The ground crew made a very significant contribution to the squadron's success. They were required to work flat out: setting up the squadron; producing sufficient serviceable aircraft to support the very intensive training and trials programmes; carrying out a never-ending series of modifications; receiving, storing and modifying the mines; and generally providing the back-up essential to enable the squadron to meet the deadlines it had been set. As in the air, Gibson insisted on the highest standards on the ground. Those who failed to achieve them quickly departed.

On 26 March, Gibson went in secrecy to visit Wallis for his initial briefing, although he could only be given limited information, since his security clearance had not yet come through. On this day Gibson was told by his RAF superior that the operation would be conducted at a very low level, with the final approach to the target being at a very precise height and speed over water, which led Gibson to speculate that the intended target was the German battleship *Tirpitz*, which was then based in a fjord in Norway.

The key to the whole operation was Barnes Wallis's mine, a large drum-shaped device 60 inches long, 50 inches in diameter, and weighing 9,250lb, which was mounted transversely between two brackets that sprang apart to release it. It had to be spinning backwards when it hit the water, so it was spun up to 500r.p.m. in its mounting before release, driven by a sprocket and chain from a hydraulic motor bolted to the cabin floor. At the time 617 Squadron was formed, the mine was still under development and it only passed its final tests days before the operation took place.

The conditions in which the mine was to be dropped were extremely demanding, requiring a precise speed (220m.p.h.), height (60 feet), distance from the target and aircraft attitude. The correct speed was a matter for the pilot, but reaching and maintaining the necessary height was more difficult. The problem was solved by fitting two spotlights, one at each end of the underside of the aircraft, so that the two circles of light just intersected at the required height. The navigator, watching for the intersection, would give oral corrections to the pilot. This solution was conceived, developed, tested and modified in April, and all aircraft were fitted with the definitive version by 9 May. The question of distance from the target at the

Möhne and Eder dams was solved for most crews by a simple triangular sight, consisting of a wooden base, a peephole and two nails, the geometry of the device being such that when the bomb-aimer found that the two towers either side of the dam coincided with the nails, he pressed the bomb release. Other crews, however, developed their own systems. Finally, the cylindrical mine had to be exactly level when it hit the water, since if it was not, it would deviate in the direction of the cant. Again this depended on the pilot holding the aircraft level at the moment of dropping.

Installation of the mine required a specially modified version of the Lancaster bomber, known as Type 464, of which twenty-three were built. Three were used for trials and the rest were delivered to the squadron, the first arriving at Scampton on 8 April, the last on 13 May. During training one of the squadron machines was severely damaged on 13 May and had to be replaced by one of the trial machines on 16 May, the day of the operation. Among many other more minor concerns, the usual high-frequency radios were found not to work at such low altitudes, and fighter VHF radios had to be obtained and fitted.

On 28 March, Gibson flew a Lancaster over a reservoir to test flying at a precise height, which proved to be practicable in daylight but extremely hazardous at dusk. Then, on 29 March, he was let fully into the secret as to the intended target. Meanwhile, a weeding-out process of the aircrews had already started. Two crews were returned to their squadrons in early April. They had not come up to the standards Gibson required. When he then wanted to return an individual navigator for the same reason, the pilot and the remainder of the crew objected, so he sent the whole lot back. In another case, an individual officer was returned to his unit on medical grounds. By 16 April the crews had been reduced to 21 (147 men) and by 24 April to a final total of 19 (133 men). Some of the ground crew had also been weeded out and returned to their original units.

By the end of April over 1,000 hours had been flown in training flights, with particular emphasis on low-level flying and navigation at night, and in crossing water at 150 feet. This process was aided when two aircraft were fitted with special screens on the cockpit windows which, in combination with special goggles, enabled simulated night flights to be carried out in daylight.

The decision to launch the raid was taken at the highest possible level – by the Combined Chiefs of Staff – on 14 May. On the same day a full squadron rehearsal was carried out, with complete success. That day also saw the first detailed briefing for the two flight commanders, Gibson's second-in-command (Hopgood) and the bombing leader (Hay). The next day was devoted to major efforts on the ground – the completion of all modifications and the loading of the weapons – although a few training flights did take place.

On 16 May the briefing of the flight commanders, pilots and navigators started at about 1200 and went on most of the afternoon, leading into a general briefing for the whole squadron at 1715. At the latter, Gibson opened the briefing, followed by Wallis, the civilian scientist, who described just how the mines worked. Then, after a few words of encouragement from the Group Commander, Gibson launched into the detailed operational plan. Afterwards there was time for a meal and some final preparations, and then the crews moved to the aircraft.

The attack, officially named Operation Chastise, took place on the night of 16/17 May and was mounted from RAF Scampton by nineteen aircraft in three waves, the first consisting of three groups of three aircraft, the second of five aircraft and the third, the reserve, also of five. All aircraft flew at an extremely low level on both the outward and the return legs to avoid enemy radar and to give flak batteries the most difficult targets. In fact, they flew so low that one hit the water, several flew under rather than over high-tension cables, and all had to take evasive action to avoid chimneys, trees and tall buildings.

Gibson led the first flight of three Lancasters of the first wave, taking off at 2139. They flew over the North Sea and crossed the Dutch coast slightly off course, but apart from some light flak when crossing the Rhine their main problem was navigation. Nevertheless, all three arrived over the Möhnesee as scheduled. After circling, Gibson then flew directly over the dam at low level to test the defence's reactions. There was considerable light flak, but no balloons or searchlights.

At this point the second flight arrived. These aircrafts had taken off at 2147 and had followed the same route as Gibson, arriving safely at 0026. The third flight took off at 2159 and followed a different route, arriving a few minutes after the second, but with only two aircraft, the third having been shot down by flak.

Gibson took firm charge of events, controlling the eight aircraft using the specially fitted VHF radios, and he started the proceedings by carrying out the initial attack himself. He later admitted to some fear during the run-in but he kept the aircraft rock-steady at the required speed and height. The mine was released at the correct point and, after skipping three times, it sank and exploded. When the water subsided it was seen that the dam had not broken, so Gibson directed the second aircraft (Hopgood) to attack. This was hit many times by flak, and the mine dropped late, so that it skipped over the dam. The aircraft, which only just cleared the top of the dam, then exploded and crashed (surprisingly, two men survived, but the other five were killed).

At 0038 the third aircraft (Martin) attacked with Gibson flying ahead to draw the flak. Again the mine was dropped correctly but still the dam wall held. Gibson then called in a fourth aircraft (Young) while he drew flak from the north side of the dam and Martin came in slightly ahead of the attacking aircraft. The attack was again accurate but still without apparent effect, so a fifth aircraft (Maltby) was called in with Gibson and Martin again orbiting near the dam to draw fire. As Maltby went in he realized that Young's mine had in fact breached the dam but, being already committed, he too dropped his mine.

The dam had indeed broken at 0056, and the seven surviving

Lancasters flew over the area, inspecting the awesome results as a huge torrent of water cascaded down the valley. Time was pressing, however, so Gibson ordered two aircraft to make for home (they arrived safely) while five Lancasters went on to the next target, the Eder dam. These consisted of two aircraft without mines, those of Gibson in command and Young, who was ready to take over if Gibson was shot down, plus the three aircraft still with mines.

Gibson was the first to find the Eder and he fired red Verey cartridges to enable the others to home in on him. This target was particularly difficult, with a steep approach, a short level run to the drop point, and then a steep climb to avoid a hill behind the target. Shannon, the first to try, made four attempts before taking a rest, following which Maudslay tried twice. Shannon then made three attempts, on the last of which he dropped the mine correctly, but without breaching the dam. Next, Maudslay tried again and released his mine but it bounced over the dam and exploded in mid-air. Maudslay did not return to Scampton, but whether his aircraft crashed at the scene or on the flight home has never been established. Third in was Knight who made one dummy run, followed by a live one, which blew a huge hole in the dam. The four surviving aircraft then set off for home and three arrived, one (Young) having been shot down off the Dutch coast.

Four of the five aircraft of the second wave had actually been the first to take off from Scampton but they had followed a more circuitous route to the dams. One aircraft (Barlow) crashed in the Netherlands, the second (Byers) disappeared without trace, a third (Munro) was so damaged by flak that it was forced to return, and the fourth (Rice) flew too low and hit the water, ripping the mine away. The aircraft itself survived but without a mine there was no point in continuing so it returned to base. The fifth crew (under McCarthy) were actually forty minutes late in taking off as their first aircraft developed a fault. However, having transferred to a stand-by aircraft, they reached the Sorpe dam safely. There, their task was to drop the mine along the line of the dam, rather than at right-angles, as at the other dams. The approach was very difficult and McCarthy made ten runs before finally dropping the mine at exactly the right point, but it caused only a small breach. He then returned to base, arriving safely at 0323.

The five aircraft of the third wave were the reserve. Their primary targets were the Möhne or Sorpe dams if these had not been breached by the first or second waves, but each also had an alternative target. They took off between 0009 and 0015 and flew singly. The first two aircraft (Ottley and Burpee) were shot down on the outward leg, but the third aircraft (Brown) reached the Sorpe two hours after McCarthy's attack and like him had to make numerous attempts before dropping the mine at 0314. It caused further damage but did not breach the dam. The fourth aircraft (Townsend) flew to the Ennepe dam, where it became the only aircraft to suffer from shuddering when the mine was spun up prior to launch. The mine was dropped at the third attempt but failed to breach the dam, and the aircraft then returned safely to Scampton, being the last to land at 0615.

The fifth aircraft (Anderson) reached Germany but was forced to take evasive action after encountering searchlights and flak and got lost. Eventually, with dawn breaking and a heavy ground mist forming, the pilot decided to return home, reaching Scampton at 0530 with the mine still on board.

Whether or not the operation achieved the anticipated strategic results is not a matter for discussion here. What is significant is that Gibson and his squadron carried out the operation in compliance with the requirements of higher command. The loss rate was high: out of 19 aircraft, 8 were lost, and of 133 men, 53 died and 3 were captured. All known casualties were due to German flak. Indeed it is astonishing that, despite operating deep within the German heartland and with eight aircraft circling over the Möhne dam for thirty minutes and five aircraft over the Sorpe for twenty minutes, not one Lancaster was ever challenged by a Luftwaffe fighter.

A large number of people worked on the project, but Gibson was the man at the centre of it all, the man upon whom success or failure depended and who was faced with a daunting set of challenges. First, he had to form a brand-new squadron from scratch, requiring the assembly of men, aircraft and equipment. Second, Wallis's mine was totally new and was being refined and tested even as the squadron formed for the operation. Third, new tactics had to be devised for the flights to and from the target area. And, finally, new tactics and equipment had to be developed for the actual delivery of the weapons, involving a very low-level approach at a precise speed, height and attitude.

Gibson had fifty-six days between reporting to Scampton and carrying out the attack, during which he had an enormous workload. He had to supervise the formation of the squadron, scrutinize and where necessary test the aircrew, and arrange for those who failed to live up to his standards to be replaced. He had to impose his standards of airmanship, discipline and technical performance, thus ensuring that the squadron performed as a cohesive entity. In addition to his activities on the base, he had to liaise with Wallis, visiting him at home on several occasions, and to attend the trials at Reculver in Kent. He had also to undertake a great deal of staff work and was deeply involved in the decision-making process throughout the preparatory period, and especially in the detailed plans for the operation.

There were also a number of diversions. High-level interest in the operation meant that he had to receive numerous visitors, including AOC 5 Group (Cochrane) on 30 April, Marshal of the RAF Lord Trenchard on 5 May, and AOC Bomber Command (Harris) on 6 May, all of which – welcome and necessary as each was – took up time and effort in general preparation, rehearsal and the actual visit. On top of that, he had to deal with complaints from the public over low flying, a task which might surely have been undertaken by someone else.

Despite his youth and the pressure of time and events, Gibson coped well. There was absolutely no doubt as to who was in command. A strict disciplinarian who set very high standards in the air and on the ground, he led from the front, both in training and on the operation itself.[4]

*

These men – two sailors, one soldier, one marine and one airman – epitomize a particular breed of commanding officer for whom war seems to be second nature. They were firm but fair with their men, and took a great deal of trouble to ensure they were well looked after, both in peace and in war. All were able to recognize that much more rested on their shoulders than the outcome of the particular engagement in which they were involved, and all were able to live comfortably with such a responsibility. Above all, however, they possessed a master's touch in the skilful deployment of their units, a quality which is always recognized by the troops under their command, giving them extra confidence and raising them to even higher levels of achievement.

3

'Warriors for the Working-Day'

It is only natural that the great epics of courage should attract the most fame and be the most frequently quoted in histories. But such moments of glory are usually of short duration. As a rule, all navies, armies and airforces depend in war upon commanding officers who continue to take their units into combat or out on patrol not only day after day but, if necessary, year after year until the conflict is resolved.

THE 2ND BATTALION, THE DEVONSHIRE REGIMENT

The Western Front, November 1914 – November 1918

During the First World War, the 2nd Battalion of the Devonshire Regiment (2/Devons) served continuously in north-west France from November 1914 to November 1918, during which time it lost 80 officers and 1,253 soldiers killed or missing. At the outbreak of war in August 1914 the battalion was in Egypt. It sailed for England on 13 September, arriving on 1 October to become part of the Eighth Division, which was being formed from regular battalions recalled from the outposts of the British Empire. Once complete, the Eighth Division moved to France. The 2/Devons arrived in the front line for the first time on 18 November, shortly afterwards carrying out its first major attack – 'going over the top' as it began to be termed – in which it lost 120 soldiers killed or missing. The men then remained in the thick of the fighting throughout the war. They fought in the epic battles of Neuve Chapelle, the Somme, Ypres and Passchendaele, and they were part of the great retreat of March 1918, but most of the time they simply endured the grinding slog of trench warfare.

The succession of commanding officers in the 2/Devons is reasonably typical of units in the First World War (see Table 2). A regular battalion of the old peacetime army, in the four years between landing in France in 1914 and the Armistice in November 1918, it had nine commanding officers. The first was Lieutenant-Colonel Travers, who took the battalion to France and remained in command until February 1916, with just one three-month break in 1915, during which he underwent a serious operation and then a brief period of recuperation. During Travers's absence his place was filled by his second-in-command, Major Ingles, who then departed to command another battalion of the Devonshire Regiment.

On Travers's departure in February 1916 he was succeeded by Lieutenant-Colonel Sunderland who, apart from two weeks' sick leave in late 1916, remained in command until 31 July 1917, when he and his adjutant were killed leading the battalion in an attack during the Third Battle of Ypres. The fourth commanding officer, Tillet, had been the adjutant in 1915–16 and stood in for Sunderland when the latter was absent on sick leave in 1916. Tillet took command when Sunderland was killed in July 1917 but was himself wounded in action at Passchendaele on 3 December that year and died two days later. Tillet had been in continuous action since August 1914, had been wounded on several occasions and was considered by the battalion to be a fine and very experienced commanding officer – at the time of his death he was 25 years old.

Tillet's replacement was Lieutenant-Colonel Imbert-Terry, who had been with the 2/Devons since November 1914, but he was removed after only two months when, according to

TABLE 2: *Commanding Officers, 2/Devons, 1914–1918*

	Appointment	Length of Command	Reason for Posting Elsewhere	Status
Travers	Nov. 1914–Feb. 1916	16 months	Posted to England	
Ingles	Mar. 1915–June 1915	3 months	Superseded	Temporary
Sunderland	Feb. 1916–Jul. 1917	18 months	Killed in action	
Tillet	Nov. 1916–Dec. 1916	3 weeks	Superseded	Temporary
Tillet	Aug. 1917–Dec. 1917	5 months	Killed in action	
Imbert-Terry	Dec. 1917–Feb. 1918	3 months	Sick	
Green	Feb. 1918–Feb. 1918	10 days	Sick	
James	Feb. 1918–Feb. 1918	1 day	Dismissed	
Cope	Feb. 1918–Apr. 1918	3 months	Superseded	Temporary
Anderson-Morshead	Apr. 1918–May 1918	1 month	Killed in action	
Cope	May 1918–Nov. 1918	6 months	Survived	

official documents, his 'health broke down'. His replacement, Major Green, lasted barely a week before his health also 'broke down' and he, too, was sent back to England. Next to be appointed was Lieutenant-Colonel James, an experienced officer, who had commanded the 9/Devons from May 1916 until March 1917, when he was wounded and invalided back to England. He was awarded the Distinguished Service Order (DSO) for this period in command, indicating that he was a reliable and successful commanding officer. However, within a couple of hours of his arrival to take command of the 2/Devons, James received an order to provide fatigue parties of 450 men per day, upon which he informed his divisional commander to his face that the Devonshire men were 'fighting soldiers and not bloody navvies'. James was dismissed on the spot.

James's dramatic and abrupt departure, the third in three months, left a yawning gap which was temporarily filled by Captain Cope, with the ranks of acting major and temporary lieutenant-colonel, who commanded from February 1918 until a new and hopefully more permanent commanding officer arrived on 7 April, whereupon Cope reverted to second-in-command. The new man, Lieutenant-Colonel Anderson-Morshead, another captain/acting major/temporary lieutenant-colonel, had gone to France in September 1914 as a lieutenant with the 2/Devons, but had transferred to the 1/Devons in September 1915. By 1917 he was second-in-command of that battalion and had to take over when the commanding officer was killed in action on 3 October 1917, remaining in post until 5 February 1918. On being relieved by a permanent commanding officer, Anderson-Morshead reverted to second-in-command of the 1/Devons, but only two months later he was posted to fill the vacancy in the 2/Devons created by James's dismissal. The older soldiers in the 2/Devons knew both Cope and Anderson-Morshead well, and it was the general assessment that both were considered 'good COs', both were men of undoubted courage and leadership, and both had the interests of their men very close to their hearts. The only difference was that Cope was held in great affection while Anderson-Morshead was greatly respected. Anderson-Morshead was, however, killed on 27 May 1918, as will be described in Chapter 5, and Cope once again resumed command and remained in post until the war's end.

The nine officers who commanded the 2/Devons between November 1914 and November 1918 were a little unusual in that they were all long-term, regular officers of the Devonshire Regiment. Many other battalions, especially later on in the war, had commanding officers posted in from totally unrelated regiments, and many officers were either commissioned from the ranks during the war or joined the army only for the duration.

Only two commanding officers, Travers and Sunderland, served for over a year, and both men had a lengthy period of sickness in the middle of their tour. Two commanding officers, Sunderland and Anderson-Morshead, were killed in action, while Tillet died two days after being wounded in battle. Three stayed only a very short time in early 1918. Of the two men whose health broke down, Imbert-Terry was awarded both the DSO and the Military Cross, while James received the DSO (nothing is known about Green), which indicates that their courage and leadership were not in dispute. It seems most probable that these were simply weary men, a suspicion borne out by James's furious row with his divisional commander on his first day with the battalion.

Between them these nine commanding officers kept the 2/Devons in the front line for four long, hard years. They ensured that the men were resolute in defence and determined in attack, and never failed in anything they were called upon to do. Indeed, as will be seen in Chapter 5, as late as May 1918 they were prepared to stand, quite literally, to the last man and the last round.[1]

KAPITÄNLEUTNANT SIEGFRIED LÜDDEN

The Far East Patrol of U188, *30 June 1943–19 June 1944*

An endurance test of a different nature was faced by another commanding officer, Kapitänleutnant Siegfried Lüdden, who was 27 years old when he set out on a voyage to the Far East in command of *U188*. The voyage was to last 354 days.

Lüdden joined the German Navy in 1936, transferred to U-boats in 1940 and, after several voyages as a watch-keeping officer, was appointed commanding officer of *U188*, a Type IXC/40 long-range U-boat. He joined the boat during its final stages of construction and, having worked up his crew, he sailed from Kiel on 4 March 1943 to operate as part of a wolf-pack in the north Atlantic. At the end of his patrol, he headed for the U-boat base in Lorient, in German-occupied France, reaching the port on 4 May 1943.

After less than two months in port, *U188* left Lorient on 30 June. Lüdden's new mission was to sail to the German U-boat base at Penang off Japanese-occupied Malaya and back, carrying out attacks in the Indian Ocean and Arabian Sea on both the outward and the return legs. He was also required to bring back a load of badly needed strategic supplies. Initially, *U188* sailed in company with *U155*, but the two boats parted on 12 July as planned. *U188* continued southwards, under instruction to rendezvous with *U487*, a Type XIV tanker, south-west of the Azores.[2]

By 18 July Lüdden had still not found *U487* so he assumed (correctly) that the tanker had been lost and asked U-boat headquarters by radio for a new plan. As a result, *U155* was ordered to abort its current mission in order to meet *U188* at a new rendezvous 500 miles south of the Azores and to refuel the Malaya-bound boat.[3] This was achieved on 22 July, when *U188* received fuel and sufficient provisions for twenty-four days. Lüdden then continued southwards, rounded the Cape of Good Hope and arrived without incident at his next rendezvous, in the Indian Ocean to the south of Mauritius, where he met a German surface resupply ship, *Brake*, on 8 September.

Lüdden next proceeded northwards and looked into Port Louis in Mauritius, but there were no targets. He carried on to the Somaliland coast where, on 21 September, he sank a US freighter, *Cornelia P. Spencer* (6,361 tons). On 27 September he chanced upon a ten-ship convoy escorted by a single destroyer but after a sixteen-hour chase he was forced to break off because of engine problems. He found the convoy again the next day and fired six electric torpedoes. The torpedoes were heard to detonate nine minutes after launch but no hits were scored, indicating a fault that was to affect all his electrical torpedoes throughout the voyage.[4] *U188* was subsequently chased by the destroyer but avoided it without trouble.

Lüdden next attacked a solitary tanker on 1 October from a range of 800 metres, again without scoring a hit, and a further attack on 5 October on a Norwegian tanker resulted only in damage, not a sinking. On leaving the Gulf of Oman, he reported by radio and was instructed to sail to Penang, which he reached without further incident on 30 October.

He remained at Penang for just over a month while the boat was scraped, cleaned and repaired, and the crew enjoyed some well-deserved rest in the hill stations of Malaya. They sailed again on 12 December, reaching Singapore the following day, where the cargo for the homeward voyage was loaded.[5]

Lüdden then returned to Penang for final preparations and set sail on 9 January 1944. He had a very successful patrol in the Arabian Sea, around the island of Socotra and in the Gulf of Aden, where he sank seven steamships and seven dhows. On 29 January he sighted several aircraft and a surface escort group but evaded them all without difficulty. On 7 February he encountered three dhows and, since they were not suitable targets for torpedoes, attacked them on the surface, halting them with gunfire and then ramming them. On 12 February he sank a further three dhows, but by 14 February he had no torpedoes left and was ordered south to rendezvous with the tanker *Brake* for a second time. He replenished from the tanker on 11 March but he was still in the area on the following morning when *Brake* was attacked. Having ascertained that the survivors had been picked up by *U168*, which was also in the area, he then resumed his course for home.

On 22 March Lüdden met the outward-bound *U1062* and received the latest Enigma keys and the boat's mail.[6] He then suffered some weather damage while making a wide swing around the Cape of Good Hope. He crossed the Equator on 22 April, where he met another outward-bound boat, *U181*, to take on board a sick officer and some desperately needed lubricating oil, and to brief the outward-bound captain on conditions in the Indian Ocean. On 28 April he made yet another rendezvous, this time with *U129*, to take over the special radar-detection devices needed for the final leg of the voyage through the Bay of Biscay.

The excitements of the voyage were not yet over, however.

On 1 May Lüdden was instructed by Admiral Dönitz, Commander-in-Chief U-boats, to resupply *U66* which was 600 miles distant and desperately short of both fuel and provisions. Lüdden accordingly altered course, estimating that he would meet *U66* on 6 or 7 April. By 5 April, however, he found that he was in an area of intense enemy anti-submarine activity. That evening he saw three destroyers in line abreast but managed to avoid them, and the next morning he observed aircraft and a carrier in the distance. He also heard radio reports from *U66* that she was being attacked by aircraft and he must have been very relieved when, at 1215, headquarters told him that the resupply had been cancelled and that he was to resume his voyage northwards.[7]

At this point *U188*'s radio transmitter packed up and for the rest of the voyage U-boat headquarters heard nothing from the boat, gradually reaching the conclusion that, like so many others, she had been lost. For Lüdden, however, the loss of his transmitter meant that the Allies could not use direction-finding (DF) equipment against him and he pushed ever onwards at the most economical speed, passing the Cape Verde and Canary Islands until he reached the Portuguese coast. There he lay on the bottom through the full-moon period until he was able to cross the final stretch of the Bay of Biscay and proceed up the River Gironde, tying up in Bordeaux harbour at precisely 1100 on 19 June. His unexpected arrival caused great surprise in U-boat command.

Lüdden's circumstances were quite different from those of the commanding officers of an infantry battalion such as the 2/Devons, since, apart from two very brief periods ashore in Penang and Singapore, he was alone, with no one to turn to for advice or help. During his voyage he sank eight merchant ships (see Table 3) and seven dhows, and was close to both *Brake* when she was destroyed in the Indian Ocean and *U66* when she was sunk in the Atlantic. Despite all that, *U188* was only once seriously harassed by enemy anti-submarine forces, she lost no men and she returned with both boat and cargo intact. *U188* also made six successful surface rendezvous, each of them

TABLE 3: *Ships Sunk by U188*

Date	Victim	Tonnage
21 Sep. 43	*Cornelia P. Spencer* (USA)	6,361
20 Jan. 44	*Fort Buckingham* (UK)	7,122
25 Jan. 44	*Fort la Maune* (UK)	8,422
26 Jan. 44	*Samouri* (UK)	7,219
26 Jan. 44	*Surada* (UK)	5,427
29 Jan. 44	*Olga M. Emibiricos* (Greece)	4,677
5 Feb. 44	*Chun Chang* (China)	7,176
9 Feb. 44	*Viva* (Norway)	3,798

fraught with peril in view of the Allies' complete air superiority and their ability to intercept and decrypt virtually all U-boat radio traffic (although Lüdden was, of course, unaware of this).

The voyage was by no means trouble-free: the electrical torpedoes proved unreliable, the diesel engines were troublesome, and the lubricating oil supplied by the Japanese in Penang was of very low quality. In addition, by the time the boat returned to Bordeaux its battery was in such a bad state that when the port was evacuated six weeks later the boat was declared unfit for the voyage to Norway and had to be scuttled. The crew found the voyage extremely tiring, especially the period in the Arabian Gulf, and Lüdden recommended that in future operational U-boats should proceed from France direct to Penang where they would refit and rest before undertaking a tropical operation.

Lüdden brought *U188* back to France 354 days after first setting out, during which time he had been at sea for 297 days and had covered 39,792 miles, 2,866 of them submerged. He had no disciplinary problems and he kept his crew effective throughout a very trying experience, all having to work extremely hard, whether it was the watchkeepers on the bridge, the radiomen on their electronic watch, the mechanics coaxing the troublesome diesels, the torpedomen rectifying the faults of their 'fish' or the single cook feeding every crew member three times a day.

The survival of his boat during the return voyage may well

have been due to prudent use of his radio transmitter, and the 'break-down' during the final stage may well not have been accidental. As for Lüdden himself, he had to remain alert and on duty every single day of the long voyage. It was by any standards a protracted and testing experience for a commanding officer.[8]

KORVETTEN-KAPITÄN SIEGFRIED, FREIHERR VON FORSTNER

Atlantic Battles, October 1941–October 1943

U402 was a Type VIIC, the class which formed the bulk of the German U-boat force during the Second World War, and while there was no such person as an 'average' U-boat commander, her captain, Korvetten-Kapitän Siegfried, Freiherr (baron) von Forstner, was typical of the best commanding officers. He was brave and dedicated, fought for three and a half years in a most demanding and relentless environment and, within the context of a hard-fought and unforgiving campaign, conducted himself honestly and decently.

Born in 1909, von Forstner was commissioned in October 1934 and served initially aboard surface warships, not transferring to U-boats until 1940. He carried out his obligatory 'commanding officer under instruction' patrol aboard *U99*, a Type VIIB. Her captain was the redoubtable ace Otto Kretschmer who, in the course of a twenty-two day patrol, demonstrated his skill to von Forstner by sinking six ships. The patrol was relatively brief, but it was packed with incident and doubtless gave von Forstner much to think about.

Von Forstner's first command was *U59*, a Type IIC in the training flotilla, in which he served from November 1940 to April 1941, but he was then given command of the brand-new *U402*, which was commissioned on 21 May 1941. Following acceptance trials, he took his boat to the training flotilla for working-up.

TABLE 4: U402 *Patrols*

	Area	Patrols			Breaks	Victims	
		Start	*End*	*Days*	*Days*	*Sunk*	*Damaged*
1	N. Atlantic	26 Oct. 41	9 Dec. 41	44	34		
2	Azores	11 Jan. 42	11 Feb. 42	32	44		1
3	US coast	26 Mar. 42	20 May 42	56	27	3	
4	US coast	16 Jun. 42	5 Aug. 42	51	60		
5	N. Atlantic	4 Oct. 42	20 Nov. 42	48	55	4	1
6	N. Atlantic	14 Jan. 43	23 Feb. 43	41	57	6	1
7	N. Atlantic	21 Apr. 43	26 May 43	36	92	2	
8A	Bay of Biscay	26 Aug. 43	27 Aug. 43	1	8		
8B	N. Atlantic	4 Sep. 43	13 Oct. 43	40			
	Totals			349	377	15	3

U402 sailed from Kiel on 26 October 1941 and headed across the North Sea, around Scotland and into the waters to the south of Iceland. She was then ordered by Commander-in-Chief U-boats to join three wolf-packs in turn but without any success. By now short of fuel, von Forstner headed for Saint Nazaire, arriving on 9 December 1941.

The boat spent just over a month in port, including both Christmas and New Year, sailing again on 11 January 1942 and heading southwards for the Gibraltar-Azores area. Von Forstner found a troop-carrying convoy north-east of the Azores on 16 January and attacked the *Llangibby Castle*, scoring one hit which demolished the rudder and killed twenty-six, but the damaged troopship managed to enter the neutral port of Horta for repairs. Von Forstner waited outside, where he was joined by *U581*, but when the troopship sailed, British escorts sank the second U-boat and prevented von Forstner from scoring another hit. That apart, the patrol achieved little and he returned to Saint Nazaire on 11 February 1942.

Six weeks later, on 26 March 1942, von Forstner sailed on his third patrol, his boat being part of the fifth wave of Dönitz's attacks against the United States, and on 13 April he sank his first victim, *Empire Progress*, 550 miles south of Newfoundland.

He then operated off Cape Hatteras for several weeks but without success, so he moved further south to the North Carolina coast where he sank the Russian ship, *Ashkabad*, on 30 April. On 2 May he encountered USS *Cythera*, a former yacht that had been requisitioned and converted into an anti-submarine patrol vessel. One of *U402*'s torpedoes caused an enormous explosion on board the *Cythera* (apparently setting off its depth-charges) in which sixty-nine men died, although two survived to be rescued by von Forstner and taken back to France, the first US navy prisoners-of-war since hostilities had begun. *U402* arrived back in Saint Nazaire on 20 May.

The U-boat sailed on her fourth patrol on 16 June and en route to the United States was refuelled in the central North Atlantic by *U460*, a Type XIV U-tanker on its first voyage. Von Forstner patrolled off Cape Hatteras again, but by now US anti-submarine forces were much stronger and better organized, and on 14 July he was attacked by Coast Guard aircraft, sustaining considerable damage from depth-charges, including a battery explosion. However, he managed to escape and, having carried out some repairs, he arrived safely at La Pallice in France on 5 August.

After two months' rest and refit, *U402* sailed again on 4 October for operations off the Irish coast. On 16 October von Forstner received orders to join other boats in the search for convoy ON137, but the weather worsened and, having found nothing, he was ordered westwards to join a new patrol line 400 miles east of Newfoundland. This group found and attacked convoy SC107 off the North American coast but the U-boats were initially driven off by the escorts. Von Forstner returned to the attack on the night of 2/3 November and achieved an astonishing success, attacking seven ships in the space of four hours, sinking four and damaging a fifth which was later finished off by another U-boat. Further attacks were frustrated by US aircraft so von Forstner refuelled from *U117* north-west of the Azores and reached La Pallice on 20 November 1942.

After their second Christmas and New Year in France, von Forstner and his crew sailed again on 14 January 1943 and joined a wolf-pack off the Irish coast, but after seven days without success the group was dispersed and *U402* then joined another group searching for a slow convoy. At midday on 4 February one of the group spotted a sixty-strong convoy heading eastwards and sent a sighting report which, although it lasted no more than twenty seconds, was just long enough for two British escorts to make a direction-finding 'fix'. They then raced down the bearing, found *U187* and so damaged her that she was forced to surface and the crew to abandon ship. Nevertheless, *U187*'s signal had been received at U-boat headquarters, and the U-boats converged on the convoy for one of the most ferocious battles of the campaign. In his first attack von Forstner sank one ship at 0347, a second at 0352, and then another three in his second attack, at 0636, 0659 and 0735 respectively. He also damaged a sixth, which was later sunk by further hits from other U-boats. In the early hours of 8 January he attacked yet again, sinking one more ship with his last torpedo.

The Allies had, however, rushed reinforcements to the scene – two more US destroyers and a US Coast Guard cutter, together with a number of aircraft – and von Forstner was surprised on the surface and forced to dive. His boat was then subjected to seven separate depth-charge attacks, although it survived and von Forstner was able to nurse it back to La Pallice, which was reached on 23 February 1943. In the battle of SC118 twelve Allied merchant ships were sunk – of which von Forstner accounted for six – at the cost of three U-boats sunk and four damaged. On his return von Forstner was awarded the *Ritterkreuz des Eisernen Kreuzes* (Knight's Cross of the Iron Cross), the seventy-seventh U-boat man to receive that decoration.

The seventh patrol started on 21 April 1943. Von Forstner crossed the Atlantic once again to join a group off Newfoundland, but although a convoy was found, dense fog prevented any attacks and the operation was abandoned on 6 May. Next day, *U402* joined a new group which eventually found a convoy north-west of the Azores. On the night of 11 May, von Forstner sank two more ships but was then driven off

TABLE 5: U402 *Combat Successes*

	Patrol	*Date*	*Victim*	*Tonnage*
Sunk	3	13 Apr. 1942	*Empire Progress* (UK)	5,249
		30 Apr. 1942	*Ashkabad* (USSR)	5,284
		2 May 1942	USS *Cythera* (USA)	602
	5	2 Nov. 1942	*Dalcroy* (UK)	4,558
		2 Nov. 1942	*Rinos* (Greece)	4,649
		2 Nov. 1942	*Empire Leopard* (UK)	5,676
		2 Nov. 1942	*Empire Antelope* (UK)	4,945
	6	7 Feb. 1943	*Toward* (UK)	1,571
		7 Feb. 1943	*Robert E. Hopkins* (USA)	6,625
		7 Feb. 1943	*Afrika* (UK)	8,597
		7 Feb. 1943	*Henry R. Mallory* (USA)	6,063
		7 Feb. 1943	*Kalliopi* (Norway)	4,965
		8 Feb. 1943	*Newton Ash* (UK)	4,625
	7	11 May 1943	*Antigone* (UK)	4,545
		11 May 1943	*Grado* (Norway)	3,082
		Totals	15	71,036
Damaged	1	16 Jan. 1942	*Llangibby Castle* (UK)	12,053
	2	2 Nov. 1942	*Empire Sunrise* (UK)	7,459
	3	7 Feb. 1943	*Daghild* (Norway)	9,272
		Totals	18	28,784

by the convoy escorts. He refuelled from U459, a Type XIV tanker, and reached La Pallice on 26 May.

This time *U402* spent exactly three months in harbour before sailing on her eighth patrol on 26 August, but faults were found in the boat and she was back in port again the next day. The faults were quickly rectified and she sailed once more on 4 September. On 20 September she attacked a convoy south of Iceland, but fog again brought the attack to a halt. When the fog lifted two days later the U-boats decided to remain on the surface and fight it out with the Allied aircraft screening the convoy while von Forstner went to the aid of *U377* which was being attacked by a Canadian Liberator. The fog then descended again and after several more fruitless searches *U402* headed south to meet a tanker, *U488*, on 11 September. Before refuelling could commence, however, aircraft were spotted so the boats moved to a new rendezvous where, under the protection of two U-boats equipped with anti-aircraft weaponry, replenishment successfully took place the following day.

On 13 September, *U402* was proceeding on the surface when she was sighted by an Avenger torpedo-bomber from the carrier USS *Card*, but instead of diving von Forstner decided to remain on the surface to fight it out. At first his anti-aircraft guns forced the Avenger to stay out of range. However, the plane had called for assistance and it was soon joined by a Wildcat fighter, leaving von Forstner with no option but to dive. This was, however, exactly what the Avenger pilot wanted, and he dashed in to drop a homing torpedo which hit *U402* less than a minute later. The U-boat sank with all hands. Fifty men died, including the intrepid von Forstner.

Von Forstner and *U402* operated for two years and five months, one of the longest continuous stints by one boat under the same captain, during which time they carried out eight war patrols in the Atlantic, spending 349 days at sea. The longest patrol was the third – fifty-six days – but apart from the one-day abortive trip in August 1943 all were of over a month's duration. During that time they sank fourteen merchant ships and one armed yacht (a total of 71,036 tons) and damaged another three, two of which were subsequently sunk by other U-boats. Von Forstner was in the thick of the trade war and, with the exception of the second, all his patrols were in the North Atlantic, operating as one of Dönitz's wolf-packs against Allied convoys. He also refuelled at sea three times, on each occasion with complete success.

Von Forstner was neither the most successful of U-boat commanders – his score fell far short of Kretschmer's forty-four ships (266,629 tons) – nor was he as famous as the 'stars' like Prien, who sank the British battleship HMS *Royal Oak* in Scapa Flow in 1939. But he motivated his crew and demanded of them the same high standards that he set for himself. He was undoubtedly a canny commander who kept away from Allied anti-submarine ships and aircraft as much as possible. He and his men were perfectly well aware of the appalling losses among their comrades in the U-boat service. They heard final radio calls from some boats while they were at sea, and the gossip ashore told of others who had simply disappeared. Nevertheless, it was because captains like von Forstner and U-boats like *U402* kept on going to sea, seldom faltering, always on the lookout for targets, and living through seemingly endless days of boredom interspersed with occasional periods of high excitement and terror, that the German U-boat campaign was sustained for so long.[9]

103 SQUADRON, BOMBER COMMAND

1939–1945

The commanding officer of an RAF bomber squadron during the Second World War faced a unique combination of challenges. Some were similar to those experienced by his counterparts in the Navy or Army, but others were completely different. Among the latter were the constant loss of life among the aircrew, the manner in which the aircrew lived in an essentially peacetime environment but were exposed to violent action and acute danger on about fifteen to twenty occasions each month, and the fact that the ground personnel, who formed the majority of the squadron, were seldom exposed to the same risks as the aircrews and certainly not the same death rates.

Within Bomber Command the basic entity to which men and women gave their loyalty was the squadron, 103 Squadron being a typical front-line unit. This unit was continuously operational from 3 August 1939 to 8 May 1945, during which time it lost 996 men killed or missing, operated four different types of aircraft and was led by thirteen commanding officers. It started the war as a light bomber squadron, equipped with single-engined Fairey Battles, with which it was deployed to

France in 1939. It returned to England in July 1940, where it became a heavy bomber unit, being re-equipped with twin-engined Wellingtons in December 1940 and retaining them until March 1942, when it converted to four-engined Halifaxes. Only six months later, however, Bomber Command decided to concentrate all Halifaxes in 4 Group, so 103 Squadron re-equipped yet again, this time with Lancasters, which it then operated for the remainder of the war.

As a strategic bomber unit, 103 Squadron's primary role was in carrying out night attacks on targets in north-west Europe. Thus, its crews attacked major cities such as Berlin, Bremen, Dortmund, Duisberg, Düsseldorf, Essen, Hamburg, Kiel, Cologne, Munich and Stettin, as well as industrial towns such as Bochum, Krefeld, Leipzig, Mühlheim, Schweinfurt and Wilhelmshaven. It also took part in the raid on the missile experimental site at Peenemünde on the Baltic coast. Some targets were even further afield, including Czechoslovakia (the

TABLE 6: *103 Squadron, RAF: Killed and Missing Presumed Killed, 1939–1945*

	Year 1	Year 2	Year 3	Year 4	Year 5	Year 6
	3 Sep. 39	*3 Sep. 40*	*3 Sep. 41*	*3 Sep. 42*	*3 Sep. 43*	*3 Sept. 44*
	–	–	–	–	–	–
	2 Sep. 40	*2 Sep. 41*	*2 Sep. 42*	*2 Sep. 43*	*2 Sep. 44*	*8 May 45*
September		3	16	8	12	2
October				26	48	1
November			1		28	5
December				32	35	18
January			6	20	18	24
February		5	1	22	21	12
March	1	1	7	36	22	44
April			11	27	19	9
May	25	3	13	3	67	
June	2	11	26	40	14	
July		12	7	59	62	
August	1	9	21	42	39	
Totals	29	44	109	315	385	115
Grand total				997		

Skoda works at Pilsen) and Italy (La Spezia, Milan and Turin), round trips of 1,800 miles that lasted eleven hours and included two crossings of the Alps. Other targets were in France, including airfields, the Atlantic ports (Lorient and St Nazaire) and, particularly in mid-1944, targets in and around the Normandy beachhead. Its final operation was on 25 April 1945 against the SS barracks in the town of Berchtesgaden, which housed the men who guarded Hitler's retreat higher up in the mountain at the Adlershof, which was attacked by other aircraft.

Most raids involved bombing land targets using a mixture of high explosive and incendiaries, but there were also frequent sea-mining operations, known as 'gardening'. These involved a small number of aircraft each delivering six 1,500lb mines, which were dropped from 10,000 to 12,000 feet using H_2S radar. Other missions included meteorological flights and the dropping of propaganda leaflets. Sometimes a more experienced crew was allowed to carry out an individual operation against a target it had itself selected. Squadron-Leader Charlesworth, for example, attacked the town of Leer on 20 March 1943, dropping ten 1,000lb bombs, but when he tried to repeat his success against a different target on 1 April he failed to return, and was believed to have crashed into the North Sea.

All Bomber Command aircrew were volunteers, their official tour on a front-line squadron being officially set at 200 hours. In practice this was interpreted as thirty operations, which usually took about five months to achieve. Having attained this magic figure the man concerned was immediately posted to a training unit or some other second-line job for six months, following which he usually returned to a front-line squadron for another tour.

The aircrew on 103 Squadron came from many countries, including Argentina, Australia, Belgium, Canada, Great Britain, Jamaica, New Zealand, the Seychelles and Yugoslavia. The seven-man crews were formed during training and then tended to stick together throughout an operational tour, although when members were killed or wounded, they obviously had to be replaced on an individual basis. Losses were heavy and constant. Bomber Command lost a total of 47,130 aircrew during the war. Over the period 3 September 1943 to 2 September 1944 alone, 16,483 were lost. And in 1942 any one individual had a 44 per cent chance of surviving his first tour, but only a 20 per cent chance of completing his second.

In 103 Squadron the loss rate was of daily significance to individuals, and the monthly figures are shown in Table 6. The squadron's total losses were 996 over the entire war, the worst period being the fifth year (September 1943 to September 1944) when 385 were killed. To look at it another way, when in April 1943 one particular crew completed thirty missions, it was the first in 103 Squadron to have done so for almost a year. (It should be noted that these are figures for confirmed and assumed dead and do not include those who became prisoners-of-war.) Some relief was given by regular leave (six days' leave every six weeks) but many later reported that after three days or so they were simply eager to get back to the unit.

When crews failed to return, the cause was sometimes known – their fate might have been seen by another aircraft, or the aircraft might have reported by radio before crashing. On other occasions aircraft managed to return across the Channel but then landed at another airfield because of damage, a shortage of fuel or bad weather over the squadron's own airfield. Once a time limit had been reached it had to be assumed that the aircraft had been lost, although the squadron did not know whether the crew had baled out and, if so, whether they had been captured and become prisoners-of-war, or whether they had been rescued and sheltered by members of the Resistance. In such circumstances, their belongings were duly collected and their next-of-kin informed. This represented a constant drain, although right up to the end of the war volunteers were always available to fill the gaps.

Not all losses occurred on operations. For example, three Halifaxes were lost in training. The first crashed into the sea during firing practice (seven dead), while two stalled and dived into the ground, one on 28 July 1942 (seven dead), the other on 1 August 1942 (twelve dead). Two Lancasters were also lost on training flights in February–March 1943: in both cases the crews baled out but the pilots stayed with the aircraft and were killed in the subsequent crash.

Many of those who baled out were quickly caught by the Germans and became prisoners-of-war. Some, however, escaped. The seven-man crew of ED751 baled out over eastern France in the early hours of 6 September 1943. Three of the crew reached Switzerland, while two went to Paris from where they were moved by the Resistance to the Channel coast and taken by boat back to England. The sixth man joined the Resistance and fought with them for over twelve months until he was able to link up with the advancing Allied ground forces after D-Day. The only unlucky one was the seventh, who reached the Pyrenees before being captured. After brutal interrogation in Paris he was sent to a camp in Germany, where he survived the war.

Bomber Command was led by the legendary Air Chief Marshal Sir Arthur 'Bomber' Harris and was divided into Groups, 103 Squadron being part of No. 1 Group. The squadron was located at Elsham Wolds station in Lincolnshire, which was commanded by a group captain. He was responsible for providing common facilities such as housing, the messes, flying control, emergency services (fire engines and ambulances) and airfield security. Other station facilities included a parachute and safety equipment section, a battery-charging shop, motor transport and an armoury.

The squadron itself was commanded by a wing-commander, supported by the squadron adjutant (chief of staff). The squadron was divided into 'flights' commanded by squadron-leaders. When equipped with Wellingtons there were three flights of twelve aircraft, whereas when operating Lancasters there were three ten-aircraft flights, though this was later changed to fifteen-aircraft flights. The senior officer within each specialization in the squadron was appointed 'team leader' of his group (be he bomb-aimer, air-gunner, flight engineer or navigator), although such appointments were, of course, in addition to their normal aircrew duties. Altogether, including aircrew, ground engineers, medics, intelligence staff, drivers, armourers and administrators, a bomber squadron numbered between 700 and 800 men and women – no mean responsibility for a wing-commander.

The overall planning of each mission was carried out according to programmes developed by Bomber Command and Group headquarters, but the squadron's individual effort was very much a squadron affair, under the overall direction of the commanding officer. When the mission and the squadron's contribution to it were announced, the cycle started with the squadron intelligence officer gathering information from the relevant records, including files, maps and photographs. Details were checked as necessary with Group headquarters and with the Pathfinder Force, and meteorological reports were assembled and adjusted according to the latest information. The only people not involved at this stage were the aircrews, for whom the first intimation as to what was happening came with the posting of the crew list, which was usually put up at some point during the morning.

The first to know the night's destination were the navigators who had to plan their routes. Then there would be a briefing session in which full details would be announced and all necessary information issued. Once the aircraft had taken off there was little to be done on the ground, apart from waiting for the aircraft to return. Then the staff at squadron headquarters had to sort out who had returned, who had been shot down and where others might be, since aircraft were often forced to divert to other stations.

Bad weather, which might include fog, low cloud or heavy rain, and which could be either over the base, en route or over the target, had a major effect on operations, sometimes stopping them altogether. On such occasions it was one of the commanding officer's responsibilities to ensure that his aircrews were gainfully occupied, and training could include cross-country flights, gunnery practice, physical training, drill or lectures on the latest techniques or equipment.

The aircrew were the visible face of the squadron, but it was the men and women of the ground elements who looked after them, repaired and prepared the aircraft, and provided the general facilities on the station. Ground crew got ten days' leave four times a year.

Front-line maintenance, servicing and the repair of each aircraft were carried out by a ground crew, led by the crew chief, and they did most of their work in the open, regardless of the weather. Their never-ending duties included repairing battle damage, changing overworked engines, and fitting, replacing and modifying equipment. On several occasions the engineers of 103 Squadron managed to get twenty-seven out of the squadron's thirty aircraft fully serviceable for an operation, which was no mean achievement.

Any more extensive maintenance or repairs were the responsibility of the squadron engineering officer and his men, who worked in a large hangar. They carried out quite major repairs, including, on one occasion in 103 Squadron, creating one operational aircraft from two damaged ones by mating the sound front end of one to the sound rear end of the other. The aircraft manufacturer, Avro, also had a team at Elsham Wolds, which carried out base repairs.

The ground element included many specializations. The armourers maintained the squadron bomb dump and bombed up the aircraft according to schedules produced by squadron headquarters. The armourers were also responsible for the machine-guns, and for removing bombs from returned aircraft which had either failed to release or had been deliberately brought back (for example, if an attack had been cancelled).

The men and women of the mechanical transport section drove all vehicles, including the buses that transported crews to and from aircraft, ammunition and bomb vehicles, fuel bowsers, ambulances, fire engines and even sanitary wagons. One of their most important and delicate jobs was to collect bombs from dumps in the area and bring them to the squadron dump on the station and thence to the aircraft. Other specializations included communications personnel, flight controllers, electricians, cooks, clerks and general administrators. The ground crews were also responsible for ground defence and the provision of aircraft guards.

During operations there was a great deal of concurrent activity among these ground personnel, with the engineers preparing the aircraft for the mission, armourers preparing and loading the bombs, tankers fuelling the aircraft, and mess staff cooking pre-flight meals and preparing in-flight ration packs.

The station also included numerous facilities for the welfare of both air and ground crews, such as a library, officers' and sergeants' messes, and canteens for the junior ranks. Of particular importance was the chaplain and the station chapel, both of which had a special role to play.

From September 1939 to May 1945 103 Squadron remained in the front line, and it sent its aircrew into battle on average three nights a week from the time it was equipped with Wellingtons in 1942 until the end of the war. Some operations were harder or more dangerous than others, and some periods were later given titles such as the Battle of Berlin and the Battle of the Ruhr but, in effect, the entire period was one continuous battle. However, an outline of the events of May 1944, the worst month in the unit's history, will give some indication of what was endured, for in that month (the month before D-Day) twelve aircraft were lost, taking with them eighty-four aircrew.

The month started well with 1 Group attacking a factory in Lyon on 1 May in which 103 Squadron's contribution was six Lancasters. The weather was excellent, the target was hit and all aircraft returned safely. The next operation was on 3 May, this time against a large German army facility at Mailly-le-Camp, south of Rheims, which housed a tank-training centre and the headquarters of the 21st Panzer Division. Fourteen 103 Squadron aircraft took part, operating at 7,000–8,000 feet and many enemy fighters were encountered. Three of the squadron's aircraft were shot down, one of them being flown by a flight commander, for a loss of twenty-one aircrew.

This was followed by a three-day stand-down, during which ground training and maintenance were undertaken and a new commanding officer, Wing-Commander Goodman, took over the squadron. The next operation was on the night of 6 May, when three aircraft took part with forty-seven from other squadrons in an attack on an ammunition dump at Aubigne-Racan, north-west of Arras. The weather was good, the bombing was accurate, no enemy fighters were encountered and all aircraft returned safely.

The following night (7 May) the airfield at Rennes St Jacques was attacked by seven aircraft. Again the weather was clear, the target was hit and there were no losses. There were no operations on 8 May, but on 9 May a coastal battery at Mardyk in Belgium was attacked without loss.

On 10 May came the first 'gardening' operation of the month, with six Lancasters each dropping six 1,500lb mines from a height of 12,000 feet, using radar. All aircraft returned to base.

The next day was not a good one for the squadron. Ten aircraft were required to take part in a raid on the railway marshalling yards at Hasselt in Belgium, as part of Bomber Command's campaign to cut the lines of communication between Germany and Normandy in the run-up to D-Day. One of the flight commanders was in hospital, so the newly arrived commanding officer, Wing-Commander Goodman, decided to take his place on his first operational flight with the unit. The cloud over Belgium was so dense that the target could not be identified and the bombers, which by that time were approaching Hasselt, received radio orders from Group headquarters to return to base. Two aircraft thought that the radio message was a German ruse and bombed another target, but eight aircraft turned back, of which two were shot down over the Netherlands (whether by flak or fighters is not known): there were no survivors from either aircraft. One of those killed was Wing-Commander Goodman. He had been in command of the squadron for just six days.

There were further 'gardening' operations on 12 and 15 May, the first involving two aircraft dropping mines in the Heligoland Bight, which was achieved without loss, while in the second, five aircraft operated against Kiel Bay, in which one aircraft was lost, although the other four saw no significant enemy activity. Three days of bad weather gave the crews a welcome respite and then on 19 May eleven aircraft took part in a successful attack on railway marshalling yards at Orléans: there were no losses.

On 21 May the target for a major raid was the industrial city of Duisburg, in which 103 Squadron's contribution was eighteen aircraft. The weather was poor, with heavy cloud, and the attack took place from a height of 22,000 feet. One aircraft was lost and another was hit by flak but managed to return to base.

On the next night (22 May) the squadron again returned to the Ruhr, this time to Dortmund. The weather was good but some fighters and flak were encountered, resulting in the loss of two more aircraft. On 23 May there were two operations. One was a 'gardening' operation, dropping mines off Aalborg in Denmark, in which three aircraft took part without loss. In the other, thirteen aircraft took part in a major attack on the western marshalling yards in Aachen. The weather was good but the defences were very strong, with many fighters and dense flak, the latter claiming one of 103 Squadron's Lancasters. The final squadron operation for the month took place on 27 May, when twelve aircraft took park in another major attack on Aachen, this time against the eastern marshalling yards: two aircraft were lost.

These activities in May 1944 involved the squadron in 133 sorties on fifteen nights, during which it dropped 621 tons of bombs and mines. Total flying time amounted to 811 hours, of which 583 were operational and 228 non-operational (that is training and aircraft test flights). Losses for the month were twelve aircraft and crews, of whom seventeen survived as prisoners-of-war and sixty-seven died, including the squadron commander and one flight commander.

The lengthy trials and tribulations of 103 Squadron were not unusual, being shared by the other units in Bomber Command, but the unit had one unique feature, the most famous of all Lancasters, ED888 (PM-M), which deserves mention as symbolizing its squadron's quality of endurance. Issued brand-new to 103 Squadron in April 1943, this aircraft made its first operational flight on 4 May and for its first twenty-five operations was flown by the same pilot and crew – in itself no mean achievement. In November 1943 the third flight of 103 Squadron split away to become 576 Squadron, though still based at Elsham Wolds, and it took ED888 with it. However, the aircraft was returned to 103 Squadron in October 1944 and remained with it until 28 February 1945, by which time it had made 140 operational flights, all of them from RAF Elsham Wolds, and 97 of them over Germany, the highest total of any Lancaster in Bomber Command. Among many adventures, it returned from one raid with no radio, both outboard engines closed down due to battle damage, and its rear turret missing. Its fuselage code with 103 Squadron was PM-M and it was known to its aircrews as 'M for Mother' and to its ground crew as 'the Mother of them all'. Sadly, owing to error or oversight, this historic aircraft was scrapped in 1947.

The men who tied all this together were the commanding officers, of whom the squadron had twelve, plus one standing in for two weeks only (see Table 7). These men's primary responsibility was to ensure that the squadron remained fully operational at all times, which meant that they had to generate the necessary number of fully trained crews and serviceable aircraft as and when required by Bomber Command and 1 Group battle plans, and this in the face of heavy losses. There is no evidence that, apart from a few individuals, any squadron in Bomber Command ever faltered in this duty.

Another challenge arose in the contrast between the lives of the aircrews and those of the ground crews. In a warship or in an army front-line unit such as a tank regiment or an infantry battalion, the entire unit, regardless of rank or specialization, goes to war together, all face a common danger, and all are operational for protracted periods. In Bomber Command it was up to the commanding officer to meld aircrews and ground crews together and to ensure that their mutual dependence was clearly understood and appreciated.

One of the personal decisions that had to be made by each commanding officer was how frequently he himself went on

TABLE 7: *103 Squadron, RAF: Commanding Officers*

Dates		Names	Departure	Time in Post (months)
From	To			
27 Jan. 39	12 Mar. 40	Wg/Cdr. Gemmel	Posted	6
12 Mar. 40	23 Nov. 40	Wg/Cdr. Dickens	Posted	9
23 Nov. 40	5 Dec. 40	Sqn/Ldr. Tait	Temporary	0.6
5 Dec. 40	30 Mar. 41	Wg/Cdr. Littler	Killed on operations	4
4 Apr. 41	25 Aug. 41	Wg/Cdr. Lowe	Posted	5
25 Aug. 41	9 Mar. 42	Wg/Cdr. Ryan	Posted	7
9 Mar. 42	9 Sep. 42	Wg/Cdr. du Boulay	Posted	5
9 Sep. 42	17 Apr. 43	Wg/Cdr. Carter	Posted	8
17 Apr. 43	18 Oct. 43	Wg/Cdr. Slater	Posted	7
18 Oct. 43	6 May 44	Wg/Cdr. Nelson	Posted	6
6 May 44	12 May 44	Wg/Cdr. Goodman	Killed on operations	0.3
12 May 44	1 Dec. 44	Wg/Cdr. St John	Posted	7
1 Dec. 44	7 Aug. 45	Wg/Cdr. Macdonald	Posted	9

operations. Some had their own aircraft and crew, and flew frequently, as did Guy Gibson, for example. Others, like the commanding officers of 103 Squadron, did not have their own aircraft and flew only on a major operation, to keep their hand in or fill in when a pilot was missing. The balance was a fine one, since too much time in the air demonstrated leadership to the aircrews but left a gap on the ground, while too much time on the ground had the opposite effect. All of 103 Squadron's commanding officers during the war took part in occasional operations, and two of them died in operational crashes. Wing-Commander Littler failed to return from Brest on 30 March 1940, and, as described above, Wing-Commander Goodman died during a raid on Aachen on 12 May 1944.

As any good commanding officer appreciated, the strain on individual aircrew was immense for they were well aware of how slim their chances of survival were. After most operations, they saw their comrades' empty beds and watched their belongings being removed – fourteen on the morning of 12 May

1944, for example. That this was a strain cannot be doubted and the 103 Squadron chaplain recorded that on numerous occasions individuals came to see him for a quiet chat and to share their fears with him. The chaplain also often found figures kneeling in silence in the chapel, obviously seeking help. In Bomber Command as a whole, the policy was that if a man 'cracked up' – and it is scarcely surprising that a few did so – he was deemed to be suffering from the unfortunately named 'lacking in moral fibre' (LMF) and was quietly removed; but there is no evidence that this ever happened in 103 Squadron.

Discipline was an area where the commanding officer had to tread carefully. His aircrew were fit and healthy young men with a dangerous job and a low life expectancy and, as a result, when not on operation they tended to live life to the full, sometimes riotously so. It required a deft touch to allow them to let off steam without overstepping the mark, but on the other hand, too lax a regime led to inefficiency and misunderstandings.[10]

These commanding officers were – and their descendants in today's armed forces continue to be – men whose names seldom featured in the headlines and who were seldom recognized, even at the time, as 'aces' or 'heroes'; men who are remembered, if at all, only within their own combat community and old comrades' associations. But these were the men, regardless of which side they were on, who in the First World War repeatedly took their battalions into the trenches and 'over the top,' or in the Second World War took their submarines, frigates and corvettes on one patrol after another, or on yet another bombing raid. For many of the men in such units it was demonstrably obvious that their life expectancy was low and reduced yet further with every operation, but they kept going, and it was the commanding officer – or, in many cases, a succession of commanding officers – who enabled them to do so. Indeed, without them the system would have collapsed. In Shakespeare's words, these men were 'warriors for the working-day'.[11]

4

Two Fateful Decisions

Commanding officers frequently have to wrestle with major decisions which will affect the lives of both those under their command and others in the combat area. Sometimes there is plenty of time to reach such decisions, but on other occasions they may have to be made within minutes, perhaps even seconds. The disastrous events that occurred during the voyage of convoy KR8 from Kilindini in East Africa to Colombo in Ceylon (now Sri Lanka) in early 1944 illustrate both situations. In the first instance, the question was, on the face of it, simple: should the merchant ships in the convoy be ordered to zigzag or not? The officer responsible for making that decision decided that they should not, which was one of the major causes of the ensuing disaster. In the second incident on the same convoy, a destroyer captain was given a very positive indication that a submerged submarine was located underneath a large number of survivors swimming in the water, and he decided that it was his duty to attack.

The battle over convoy KR8 involved the captains of three British warships and one Japanese submarine, and it was one of the most unusual encounters of the Second World War. All four commanding officers were thoroughly experienced and had distinguished war records, and during the battle between the surface ships and the submarine, in which no quarter was asked and none given, the balance of advantage swung first

73

one way and then the other. Indeed, the engagement started with the British crews witnessing one of the worst maritime disasters of the war, following which one destroyer was almost sunk and the other became the only Allied warship to sink one submarine from each of the three Axis navies.

CAPTAIN JOHN WILLIAM JOSSELYN

Convoy KR8, 5–12 February 1944

The Kenyan port of Kilindini (Mombasa) was a scene of great activity during the first week of February 1944, as 6,311 passengers for the voyage to Colombo boarded five troopships, all former passenger liners – *City of Paris*, *Ekma*, *Ellenga*, *Khedive Ismail* and *Varsova*. The passengers were scheduled to join the British forces in Burma, Ceylon and India fighting the Japanese, and they included drafts and individuals from the United Kingdom as well as several army units raised in Great Britain's African colonies. What made the passenger list unusual for wartime, however, was that it included eighty-six servicewomen, plus one mother and child, all of whom, for administrative convenience, were placed aboard the *Khedive Ismail*.

The five troopships together with a naval escort were to form convoy KR8 and, following the usual practice, the 'convoy conference' took place on 4 February, the day before sailing. Those present included the senior officer of the escort, Captain J. W. Josselyn, RN, commanding the cruiser HMS *Hawkins*, and the masters of the merchant vessels, of whom Captain Whiteman of the *Khedive Ismail* was the senior and thus the *ex officio* convoy commodore.

The major issue discussed at the meeting was whether or not to zigzag – the manoeuvre by which all ships changed course at regular intervals to frustrate enemy submarine commanders. This was a tried and tested anti-submarine tactic, but it required discipline and vigilance by all watchkeepers, and it also reduced the convoy's speed of advance towards its destination by some 2 to 3 knots.[1] And therein lay the problem. The sailing orders for the convoy, issued by the Commander-in-Chief in Ceylon, required it to steam at a speed of advance of 13 knots, which also happened to be the maximum speed of the *Khedive Ismail*. Thus, the convoy had three choices: it could zigzag and arrive later than the Commander-in-Chief had required; it could ask the Commander-in-Chief for an extension of time in order to zigzag; or it could decide not to zigzag at all.

The issue was further complicated by the fact that several of the merchant captains showed a distinct lack of enthusiasm for zigzagging and in this they were supported by the convoy commodore, Captain Whiteman. The final decision on such a tactical matter rested with Captain Josselyn, a very experienced officer who had been on escort duties with no fewer than thirty-three convoys since the war had begun. He agreed not to zigzag with great reluctance, though he told Whiteman that he would bide his time and review the situation once they were at sea.

Even so, Josselyn was unable to stifle a nagging feeling of doubt and decided to consult the chief staff officer to the Flag Officer East Africa (the local naval headquarters). The staff officer told Josselyn not to worry and added that there was now a school of thought that believed not much was achieved by zigzagging anyway, since the 2-knot reduction in the speed of advance made it easier for a chasing enemy submarine to catch up. Josselyn returned to his ship half wishing to believe what he had been told but still with a nagging sense of unease.

There were strong security precautions at Kilindini before departure, but eventually all was ready and the convoy sailed in the early afternoon of 5 February, heading for Colombo, 2,700 miles away, where its estimated time of arrival was before dusk on 13 February. The principal escort was HMS *Hawkins*, an elderly cruiser completed in 1919 and designed for surface warfare, being armed with seven 7.5-inch and four 4-inch guns. However, she had no anti-submarine capability, lacking both sonar with which to locate submarines, and depth-charges with which to attack them.

Initially, the escort also included three small Royal Navy warships: a corvette, HMS *Honesty*, and two former US Coast Guard cutters, HMS *Sennen* and HMS *Lulworth*. These three ships provided limited anti-submarine protection for the first four days of the voyage, but they had only a short range and in the early hours of 9 February they were forced to leave the convoy and return to Kilindini. For the next four days the convoy had only *Hawkins* as escort. She zigzagged ahead with surface radar and visual lookouts only, which meant that the convoy, with its valuable load of troops, had no anti-submarine protection at all.

Meanwhile, Josselyn and Whiteman were wrestling with another problem. They knew that movement in and out of the port of Colombo was banned between dusk and dawn, which meant that unless they arrived in daylight, they would have to wait outside in what was obviously a vulnerable area, and particularly so in moonlight. Since leaving Kilindini, however, the convoy had experienced an unexpected adverse current of 2 knots, which reduced its speed of advance to approximately 11 knots. This meant that its estimated time of arrival had slipped from the late afternoon of 13 February to 0800 on 14 February.

The two men discussed these navigational problems by signal and also returned regularly to the vexed question of zigzagging. On 9 February, they tentatively agreed to start zigzagging at dawn on the 11th, but when that time arrived it now appeared unlikely that they would reach Colombo during daylight on the 14th even at their current speed and certainly not if they started to zigzag. Rather than spend yet another night at sea, the question of zigzagging was therefore put off for another day. At dawn on the 12th, the adverse current was still slowing them down and one of the troopships, *Varsova*, was having great trouble in keeping on station even when sailing on a steady course. However, Josselyn was under the impression that there were no enemy submarines in the area and, as a result, when Whiteman suggested that they should postpone zigzagging for yet another twenty-four hours and only begin at dawn the following morning (13 February), Josselyn replied: 'Yes, I feel we cannot afford to zigzag yet . . . Let us reconsider the matter at dawn tomorrow.'

Minutes after this exchange of signals, two destroyers of the new escort, HMS *Paladin* and *Petard*, came steaming over the horizon from Ceylon to join them. In fact there should have been a third, HMS *Penn*, but she had been prevented from sailing by engine trouble. The escort was thus smaller than planned, but in war one must frequently make do with less than adequate resources.

As the senior of the two destroyer captains, Commander Egan in *Petard* agreed by signal with Captain Josselyn that his two ships would carry out anti-submarine sweeps 3,000 yards ahead of the convoy. The destroyers were in position by 0800, but shortly afterwards Egan moved them further out still, to 8,800 yards, so that any submarine approaching on the surface could be spotted before it submerged to lie in wait for the convoy. The sea was calm, there was virtually no wind, and visibility was excellent. At 1420 two torpedoes hit *Khedive Ismail*, which sank in less than one minute with enormous loss of life.

For Captain Josselyn, the loss of the *Khedive Ismail* was the outcome of a series of decisions which, when each was considered in isolation, must have seemed relatively innocuous but which, in combination, led inexorably to disaster. A very experienced officer, he knew the Admiralty regulations well, one of which stated that the senior officer of the escort forces was 'responsible for the safety of the convoy from enemy action' and another that 'the decision as to whether zigzagging shall be carried out or not is the responsibility of the Senior Officer of the Escort'.[2] Even in the aftermath of the disaster, Josselyn never, for one moment, denied either knowledge or understanding of the regulations.

That said, the dividing line between the responsibilities of the senior officer of the escort and those of the convoy commodore was a fine one. The commodore was in charge of the

4. Convoy KR8 Dispositions at the Moment of Impact of Torpedoes on the *Khedive Ismail*, 1435, 18 February 1944

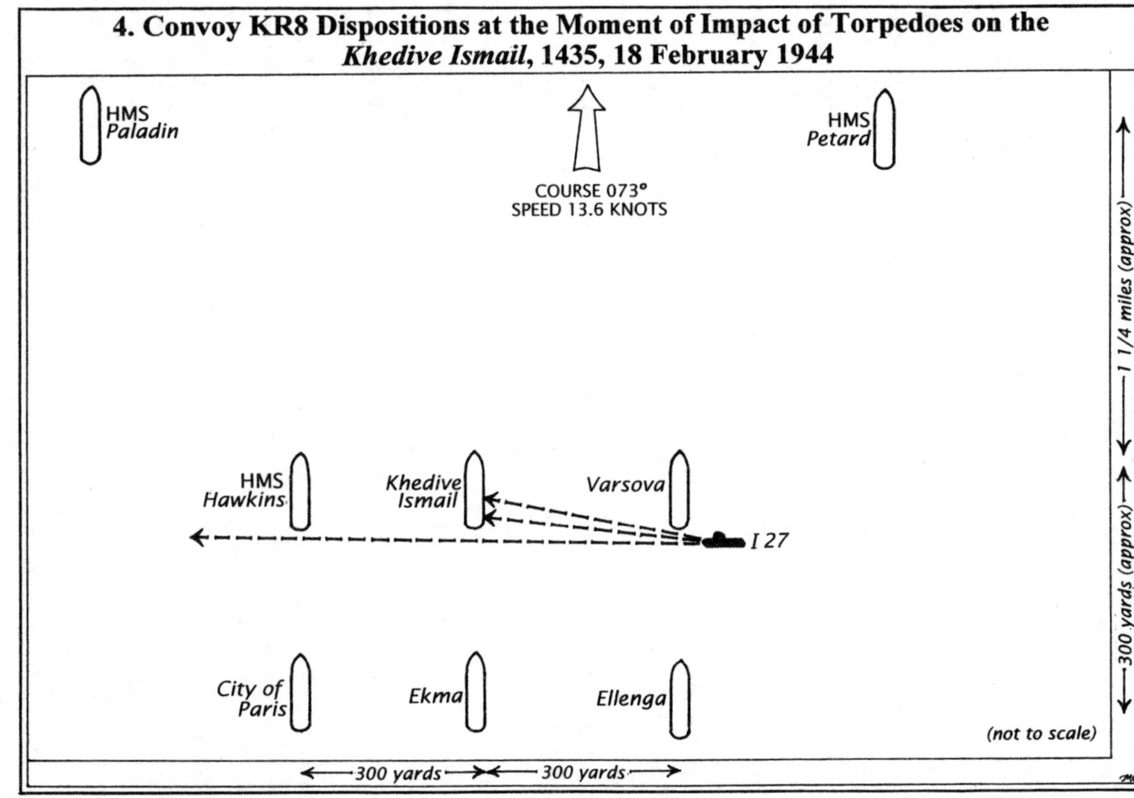

convoy, while the senior officer was responsible for the convoy's defence and safe arrival at its destination. As to zigzagging, the decision was, at least in theory, the commodore's but it was subject to the advice of the senior officer, and the regulation just quoted gave the latter full authority to insist on zigzagging if he felt it necessary to do so.

Josselyn had clearly been uncomfortable with the 'no zigzag' decision. There had been an argument at the convoy conference with Josselyn on one side and Whiteman on the other, and after the meeting Josselyn had gone to discuss the matter further with the staff officer, who fobbed him off with platitudes. Once at sea he regularly returned to the subject, but there was always some reason – the adverse current, the delay, the closure of Colombo harbour during darkness, and, finally, *Varsova*'s poor station-keeping – for putting it off.

Ironically, the Japanese submarine, *I27* found herself in the general vicinity of convoy KR8 only by pure chance. Since she was both slow-moving when submerged (with a maximum speed of 7 knots) and lacking in manoeuvrability, there was no question of her having chased a convoy that was moving at 12 to 13 knots. The convoy had had the misfortune to run right on to her. It was ironic too that the attack occurred at a time when the convoy's anti-submarine escort was at its strongest and that the huge submarine was not detected by the experienced sonar operators aboard *Petard* and *Paladin*. Had the encounter taken place at any time in the previous three days, the convoy would have had no means of detecting or attacking a submarine whatsoever. And it was sheer bad luck that Captain Fukumura, the commander of *I27*, was one of the few Japanese submarine commanders who specialized in sinking merchant ships – his 'score' prior to the *Khedive Ismail* was thirteen (75,467 tons), one of the highest in the Imperial Japanese Navy. The majority of his fellows greatly preferred to attack warships.

A Board of Enquiry into the sinking was held in Colombo on 16 and 17 February 1943 at which Captain Josselyn was questioned in great detail.

The matter of the steering of a steady course by the Convoy was carefully studied. [Captain Josselyn] came to his decision not to zigzag deliberately and after much thought and after discussion both at the conference and by signal with the Commodore of the Convoy. It is beyond question that any zigzag would have resulted in a delay of the convoy, amounting to an additional night at sea in potentially dangerous waters, and Captain Josselyn fully appreciated this, as he did that no submarine warnings were in force. In the opinion of the Board [however] his decision, made in no haste and with no negligence, was the incorrect one, placing as it did an undue reliance on the anti-submarine powers of an inadequate escort and over-stressing the additional danger of an extra night at sea.

The adverse current, which all considered unusual at that time of year, was certainly a factor outside Josselyn's control and was the initial cause of the delay. But the anti-submarine resources allocated to Captain Josselyn were woefully inadequate. His own ship had no anti-submarine capability whatsoever. The initial escort had provided only limited anti-submarine protection, and that only for the first four days of the voyage. For the next three days he had no anti-submarine escort at all, and then he only received two out of the three destroyers that had been promised.

Though it is true that the anti-submarine resources available to the Commander-in-Chief Eastern Fleet in Colombo were inadequate to cover all his commitments in the Indian Ocean, it is also true that the naval high command gave the convoy, which was carrying more than 6,000 badly needed troops, a lower priority than other deployments in the area. The transport of a floating dock (admittedly a valuable piece of *matériel* and a vulnerable target) and exercises by the main fleet were supplied with much greater protection by aircraft and surface anti-submarine ships. It is also worth noting that while another destroyer could not be found to replace the defective *Penn* on

11 February, no fewer than three destroyers were dispatched to protect the convoy after the *Khedive Ismail* had been sunk on the 12th.

The unfortunate Josselyn subsequently received a letter from the Board of Admiralty in London informing him that he had committed two errors of judgement. The first was that 'you accepted the 13 knot advance schedule for the convoy which left no time in hand for zigzagging', the second 'your decision to dispense with the elementary [anti-submarine] precaution of zigzagging to avoid extra time at sea.'[3]

Thus, like many other commanding officers before and since, Josselyn was 'damned if he did and damned if he didn't'.

COMMANDER RUPERT EGAN

Sinking IJN Submarine I27

Several hours before the attack on the *Khedive Ismail* the two British destroyers *Petard* and *Paladin* joined convoy KR8. These were both experienced ships: *Petard* (Commander R. Egan, RN) had already achieved the unusual distinction not only of

TABLE 8: *The Battle of Convoy KR8: The Naval Participants*

	HMS *Hawkins*	HMS *Paladin*/ HMS *Petard*	IJN *I27*
Type	Cruiser	Destroyer	Submarine
In service	1919	1941	1940
Displacement	12,800 tons	2,430 tons	3,654 tons (submerged)
Length	605 feet	345 feet	356 feet
Speed	29 knots	36.5 knots	23.5 knots (surfaced), 7 knots (submerged)
Guns	7 x 7.5-inch 4 x 4-inch	3 x 4-inch 3 x 40mm AA	1 x 5.5-inch, 2 x 25mm AA
Torpedo tubes	6 x 21-inch	8 x 21-inch	6 x 21-inch
Depth charges	Nil	77	–
Cew	690	210	101

sinking a German and an Italian submarine, but also of managing, in both cases, to retrieve a large amount of cryptographic material before the victims sank. Her partner in this action, *Paladin* (Lt. E.A.S. Bailey, RN) had also sunk a U-boat in the Mediterranean. Then, after an active career in the Mediterranean, the two ships were transferred to the Indian Ocean, arriving at the main naval base at Colombo in February 1944.

Immediately before leaving the Mediterranean both ships took part in operations in the Dodecanese, where they played their customary aggressive role. One consequence of this was that *Petard* used up her entire supply of semi-armour piercing ammunition (SAP), a routine event in wartime and one which nobody could ever have imagined might cause a problem. However, on arriving at Ceylon, Egan took his two ships to the ammunition jetty and discovered that there was indeed a difficulty. It transpired that P-class destroyers, of which *Petard* and *Paladin* were the first to arrive in the Far East, were unusual in that their main armament consisted of four 4-inch guns, a type normally employed in cruisers as an anti-aircrft gun, while all other destroyers in the theatre were armed with 4.7-inch guns. As a result, the ammunition depot could supply only anti-aircraft shells and not the SAP which the destroyers needed for use against surface targets. The matter was quickly forgotten when Egan received orders for a routine mission, requiring him to take *Petard* and *Paladin* to meet convoy KR8 in mid-ocean and escort it to Colombo, but it was to have a considerable bearing on subsequent events.

As described above, the calm of 12 February was shattered at 1420 by two tremendous explosions as *Khedive Ismail* was hit by two torpedoes. The troopship took an immediate and rapidly increasing list to starboard and then sank in less than a minute. Of the 1,507 people on board 1,302 were lost, including 79 of the 87 women and the only child aboard. Those who did not go down with the ship were left struggling in the water. More torpedo tracks were then seen and, although none scored a hit, Josselyn ordered the surviving troopships to disperse immediately to a prearranged rendezvous over the horizon and left *Petard* and *Paladin* to hunt and destroy the enemy submarine.

On hearing the explosion, the destroyers had turned 180 degrees outwards and had headed at top speed for the spot where the *Khedive Ismail* had gone down. *Paladin* reached the area first and began to recover survivors while *Petard*'s sonar operators quickly made contact and established that the enemy submarine was moving slowly from right to left. When this information was received on the bridge it was realized that the target was directly underneath the survivors struggling in the water – every submarine-hunter's worst nightmare. Egan rapidly assessed the situation, decided that he had no option, ordered a reduction in speed to 18 knots and commenced the attack run, ignoring shouts from some of the lookouts. (It was afterwards said by some of the crew that the eight subsequent depth-charge explosions seemed louder and stronger than usual, perhaps because their effect on the survivors struggling in the water was all too apparent).[4]

Aboard *Petard* the sonar still had a firm contact after the first run so Egan took her in for a second attack, while the horrified crew watched as they passed bodies floating lifeless in the water. Then it was time for a third run, on this occasion in co-operation with *Paladin*, who had halted her rescue operation and was now under way again. But then, to the surprise of the British, a huge grey submarine surfaced a little over 2,500 yards away and small figures were seen running along the deck to man the gun. The enemy was quickly identified as a B1 type which was armed with six torpedo tubes, a single 5.5 inch gun and two 25mm cannon. The greatest immediate danger was the 5.5 inch gun which could devastate the unarmoured destroyers. Fortunately, the destroyers' 4-inch guns quickly found the range and two complete Japanese guncrews were destroyed, following which the gun itself was hit and rendered unusable.

At this point, Lieutenant Bailey, commanding *Paladin*, con-cluded that the submarine was about to dive and, being an impetuous man noted for his instant decisions, he ordered the survivors now crowding his ship to brace themselves, called for full speed and set course for the enemy. On *Petard*'s bridge everyone was watching the effect of their gunfire on the Japanese submarine, now only 1,000 yards away, when Egan realized what *Paladin* was up to. Recalling a recent Fleet order that discouraged ramming because of the number of destroyers and corvettes damaged as a result of doing so in the Atlantic campaign, he signalled *Paladin* to cease the attack forthwith.

Aboard *Paladin* the noise of battle was deafening, and the ship was shaking as it rushed at full speed towards the submarine, but the yeoman heard *Petard*'s siren and saw the frantically flashing bridge lamp ordering: 'Negative ram.' Bailey had no option but to obey his superior officer. However, his order to the coxswain was drowned by the noise of firing and there was a brief but fatal delay. As a result the turn came too late, and *Paladin*'s speed caused her to slide outwards, slamming into the Japanese submarine broadside on. Unfortunately, the destroyer was also leaning into her turn, exposing her bottom, so that *I27*'s forward hydroplane, sticking out at right-angles from the hull, ripped through the thin plates of the destroyer's hull, gouging an 80-foot hole along the bottom. At the same time, *Paladin*'s cutter, still turned out on its davits, hit *I27*'s periscopes and bent them, thereby blinding the Japanese captain for the remainder of the engagement.

Paladin limped away, Bailey's aim having changed in a few seconds from the destruction of the enemy to the survival of his own ship, while Egan placed *Petard* between *I27* and the damaged destroyer. As *Petard* ran up the side of the submarine, her cannon and machine-guns were lowered to maximum depression to rake the deserted conning-tower and upper deck at point-blank range. In bringing his vessel so close to the submarine, Egan also hoped to be able to drop twelve depth-charges, set for shallow explosion, to damage the submarine further. This was, however, an extremely risky action since it was always possible that the Japanese commander might swing his submarine towards the British destroyer and gouge a hole in her side, as had been done to *Paladin*. That he did not do so was probably due to his inability to see what was happening through his damaged periscopes.

Egan repeated the tactic twice, albeit without any apparent result, and then was faced with a major dilemma. The 4-inch anti-aircraft ammunition was unable to penetrate the submarine's hull, while the depth-charges fuses could not be set to explode sufficiently shallowly to damage the submarine. Meanwhile, *I27* was unable to dive and her gun and periscopes were unusable. However, she still represented a threat, for she could, with her huge bulk, sink the destroyers if only her captain could get in a position to do so. It was to be presumed, also, that she still had some torpedoes on board. Finally, as the British well knew, her Japanese captain and crew would never surrender. Ironically, *Hawkins*, whose 7.5-inch shells would have penetrated the Japanese hull, was over the horizon guarding the remainder of the shaken convoy.

Boarding was one possibility, but, even if men managed to get on board, they would have to fight their way, one at a time, down the main hatch, against the determined resistance of ninety-odd Japanese. That left eight 21-inch torpedoes, mounted in two quadruple launchers, for which the standard drill was to launch a 'fan', which gave the best statistical possibility of a hit. In this, either four or eight torpedoes were launched in succession while turning, the interval between launches depending upon the target's course, speed and range, and the launching ship's radius of turn and speed.

Egan, however, ordered that the torpedoes be launched singly and, to the consternation of all in *Petard*, the first torpedo, launched at precisely 1700, missed. The calculations were redone and, with the target still moving on a constant course and speed, the second torpedo was launched at 1710; that, too, missed. So, sickeningly, did the third and fourth.

By now the men on the upper deck were in agony. They had proved their loyalty time and again since Egan had taken command the year before, but everyone who could see what was going on knew that he was making a mistake. Most kept quiet, but the torpedo officer called the captain on the telephone to remind him that the wrong procedures were being used and to request that he be allowed to take local control and finish off the enemy 'according to the book'. The captain refused but seemed unable to bring himself to launch the last four torpedoes simultaneously, and the desperate encounter continued. The tension was palpable, especially on the bridge, but naval discipline and personal loyalty to the captain prevented an open protest. So, on the captain's orders, the fifth torpedo was launched, forty minutes after the first, an action which committed them irrevocably to single shots. That, too, missed, as did the sixth. By now the tension throughout the ship was almost unbearable, and the torpedo officer, his patience exhausted, stood on the after deck, shouting at the bridge and demanding to be allowed to take over local control. He was, however, ignored and all realized that some new form of attack in this extraordinary confrontation was rapidly becoming inevitable.

They had little hope left as the seventh torpedo was launched, but then there was a huge explosion, and a column of water, flames and smoke rose above the Japanese submarine. The vessel rapidly disappeared, taking every member of its crew to the bottom.

Three hours and fifty minutes had passed since the *Khedive Ismail* had been sunk, two and a half hours since the submarine had been forced to the surface, and seventy minutes since the first torpedo had been fired. There was no elation or cheering aboard *Petard*, even though some realized that the destroyer had just become the only Allied warship to have sunk a German, an Italian and a Japanese submarine in the same war. *Petard* quickly went alongside *Paladin* to help the damaged destroyer, taking on board the still-shocked survivors, together with half *Paladin*'s crew. Then, with the damaged destroyer under tow, they set course for Addu Atoll in the Maldives, arriving the following morning.

Egan was the victim of the same series of decisions as

Josselyn, but he was also affected by other factors, not the least of which were the lack of SAP rounds and the fact that *Hawkins*, whose 7.5-inch shells would certainly have penetrated *I27*'s hull, had been compelled to attend to the convoy. Why the destroyers' sonar operators failed to find *I27* as they passed over it will never be known. Both crews were battle-hardened and thoroughly experienced in anti-submarine operations, but the science of sound detection is notoriously imprecise, especially in tropical waters and using unsophisticated equipment.

Egan's initial reactions were correct. Both escorts turned outwards and headed at full speed for the scene of the action, where *I27* was quickly detected. Whether Fukumura ran under the survivors deliberately or accidentally will also never be known, but Egan's alternatives were to prosecute a certain target and prevent it from carrying out any further attacks, which meant attacking through the survivors, or to wait for the target to clear the survivors, with the possibility that it might then either torpedo one or both of the destroyers or escape altogether to sink more ships elsewhere. He chose the former course of action and succeeded in forcing the submarine to the surface. It is likely that the weight of responsibility for such a decision weighed heavily on him, and that it may have influenced his subsequent actions.

When Bailey decided to ram the submarine, he may have been convinced that he had to take quick action, but there can be no doubt that he was over-impetuous and that he did not inform Egan of his intentions. As a result, Egan's split-second decision to cancel the ramming was probably given too late, a problem compounded by the delay experienced aboard *Paladin* in passing the order to the coxswain. The resulting damage to *Paladin* not only cut the British attacking strength by half but also meant that Egan had to protect *Paladin* at the same time as he endeavoured to find a way to sink the submarine.

Finally, there was the question of the torpedoes. Egan went against all naval doctrine in launching the torpedoes singly. He never offered an explanation as to why he did so. Perhaps, since the enemy was only 1,000 yards away and proceeding on a steady course at a constant speed of 6 knots, he believed that the submarine was a sitting target. Whatever the reason, Egan was the captain, the decision was his, and his crew obeyed him, but he was extremely fortunate that the seventh torpedo obtained the desired result, since his options, once he had run out of torpedoes, were very limited, not least because Bailey had already been forced to jettison his own torpedoes to reduce his damaged ship's topweight.

The most awe-inspiring of Egan's decisions was to attack through the survivors. There is no question that he knew precisely what he was doing – the reports of others on the bridge confirm that – and the decision must have caused him acute anguish, anguish that cannot have been eased by the obvious and understandable distress among his own upper-deck crew who could see what was happening. But, while most of the survivors from the *Khedive Ismail* mention this incident in their reports, and some describe what it was like to feel the shock waves of the depth-charges through the water, not one of them contains even the slightest hint of criticism of the captain for doing what they clearly considered to be his duty.

5

To the Last Man and the Last Round

One certain way for a commanding officer to achieve enduring military glory is for his unit to fight to the last man and the last round, three of the better known examples being the battles of the Alamo, Camerone and Little Big Horn. The first of these took place during the Texan War of Independence against Mexico, when General Antonio de Santa Anna, with 4,000 men under his command, surrounded a Texan outpost at the Alamo, just outside San Antonio, commanded by Colonel William Travis. When the siege started on 23 February 1836, Travis had 155 men. He received a further 32 reinforcements on 1 March, but these were slowly reduced as the battle progressed. The end arrived at 6 March when Santa Anna's men finally stormed the fort and the last of the garrison died.

On 30 April 1863, during the French campaign in Mexico, Captain Jean Danjou was in temporary command of the 3rd Company of the 1st Battalion, French Foreign Legion, which comprised two other officers and sixty-two men. Danjou was a veteran who had lost a hand at the Battle of Magenta in Italy in 1859. In the Mexican campaign he undertook the defence of an obscure farmstead at Camerone, and he and his men held the place against over 2,000 Mexicans for ten hours, but in the final attack, unlike at the Alamo, the Mexicans showed mercy to the last three men left alive. This small battle remains to this day at the centre of Foreign Legion mystique, and Captain

Danjou's wooden hand, which had replaced the hand lost at Magenta and which was recovered after the battle, is paraded annually at the Legion Depot on the anniversary of the battle.

During the 1876 campaign against the Sioux Indians, Lieutenant-Colonel George Custer, commanding the 7th US Cavalry, was sent forward with orders to work his way round to the south of Chief Crazy Horse's position and block his escape route, following which Major-General Terry could carry out a frontal attack. Custer, however, had his own ideas, and on 25 June 1876 he approached the Sioux position above the river known as Little Big Horn head-on. When close to the Indian position he then divided his regiment into three columns. One was sent to the left, the second to the right. Custer himself led the centre column of 212 men straight into Crazy Horse's centre. It was a foolhardy plan. Custer's column was annihilated, and the other two columns escaped only after heavy losses.

There are, however, many more examples of this type of engagement, three of which will be considered in detail here.

LIEUTENANT-COMMANDER EUGENE ESMONDE

The Attack on the German Battlecruisers, 12 February 1942

Lieutenant-Commander Eugene Esmonde, RN, commanded 825 Squadron, Fleet Air Arm, during an attack against a German naval task group that ran the gauntlet of the English Channel in February 1942. Esmonde was a descendant of a well-established Irish Catholic family, and his ancestors included an uncle who had won the Victoria Cross in the Crimea and another who was executed for treason against the British Crown. Esmonde was born in 1909 and educated at the Jesuit College in London. He then spent three years studying to be a missionary with the Mill Hill Fathers but left them in 1928, having decided that he did not have a religious vocation.

Nonetheless, he remained a serious and committed Roman Catholic.

Esmonde devoted the remainder of his life to aviation. In 1929, he joined the Royal Air Force on a short-service commission, during which he spent most of his time with the Fleet Air Arm, then provided by the RAF.[1] He left the RAF in 1933 and the following year joined Imperial Airways. By 1939 he had reached the rank of captain. Then, with war looming, he received an unsolicited invitation from the Admiralty to rejoin the Fleet Air Arm, which had been transferred to naval control. He did so on 14 April 1939, with the rank of lieutenant-commander on a fifteen-year commission. Because of the impending war, Esmonde was given no additional training but simply supplied with a uniform and left to get on with it. He commanded a training squadron from May 1939 to May 1940 and was then appointed to command 825 Squadron, equipped with Swordfish aircraft. The squadron flew from various bases on the south coast, mainly on anti-submarine patrols, although it also undertook periodic artillery-spotting duties for naval and army batteries firing on German targets on the French coast.

Since Esmonde spent the remainder of his life flying Swordfish aircraft and since this unique flying machine played a key role in the events that follow, it deserves a full description. The Fairey Swordfish was a low-powered biplane, constructed from a wooden frame covered in fabric and possessing two open cockpits, one for the pilot, the other for the observer and air-gunner/radio operator. The Swordfish had entered service with the Fleet Air Arm in 1938, just at a time when every other major airforce was re-equipping with high-powered, all-metal monoplanes with closed cockpits. It was therefore an anachronism, but it was used throughout the war, from both carriers and airfields, for torpedo strikes, anti-submarine operations and minelaying. It cruised at a speed of between 70 and 85 knots. Indeed, it was so slow that modern German fighters found it very difficult to engage, while German shipborne anti-aircraft systems were designed for a minimum target speed of 100m.p.h. Its armament consisted of a single 0.303-inch Lewis machine-gun on a flexible mounting in the rear cockpit, and when the plane was employed as a torpedo-bomber it carried a single 21-inch torpedo weighing 1,610lb under the fuselage, whose weight and drag reduced its speed still further. Some aircraft carried radios for communication with the base or the carrier, but communication between aircraft was done by hand signal or by lamp. The aircraft was a delight to fly and extremely manoeuvrable, but in a torpedo-aiming run it had to fly flat and level at a constant – and very slow – speed and at a very low height, which made it extremely vulnerable. It has achieved an almost heroic status in some quarters, but a very distinguished and experienced naval pilot, who flew throughout the Second World War, took a different view:

> I would not detract from the great actions in which the Swordfish participated, nor especially from the gallant aircrew who fought these actions, but the hard fact is that these aircrew should never have been exposed to such dangers in equipment so ancient in concept and I cannot believe that a more technologically advanced aircraft could not have done as well or even better.[2]

In September 1940 Esmonde embarked his squadron aboard the aircraft-carrier HMS *Furious* for operations off the Norwegian coast. On the 22nd he led two squadrons of Swordfish in a night attack against German ships in Trondheim fjord. The weather was bad and deteriorating, and Esmonde was all for pressing home the attack, but after some heated discussions he was persuaded against it by his observer/navigator. This was followed by a period ashore, following which 825 Squadron re-embarked aboard *Furious* and flew anti-submarine patrols during a voyage which took the ship to Takoradi, thence to Gibraltar and finally home again, arriving in April 1941.

On 17 May the squadron embarked aboard HMS *Victorious* and took part in the operation against the German battleship, *Bismarck*. On 24 May Esmonde led his squadron, most of them very inexperienced, particularly in night operations at sea, in a night attack against the German battleship. Eight aircraft took part and the attack was pressed home with great determination, but it resulted in only one hit that did not inflict any serious damage.[3] Esmonde was awarded the DSO for this action. On 13 June he transferred his squadron to HMS *Ark Royal*. After taking part in various operations in the Mediterranean the squadron was still on board the carrier when she was hit by a torpedo on 13 November, 30 miles off Gibraltar. Esmonde helped in every way he could and was eventually the penultimate man to leave the ship, the captain being the last to do so. During this affair he was particularly noted for his care of his officers and sailors, both during the abandonment of the sinking ship and after their arrival in Gibraltar, and also for ensuring that any Roman Catholic who required it was given spiritual help by a priest.

In due course, Esmonde and his men returned to Britain where he set about obtaining reinforcements and working the squadron up to operational readiness once again. At this time one of the most dangerous naval threats to Britain was posed by three German warships: two battle cruisers, the *Scharnhorst* and the *Gneisenau*, and a heavy cruiser, the *Prinz Eugen*. After sorties into the Atlantic, all three had become trapped in the naval base at Brest. Hitler decided that the most effective course was to recover them to Germany and accordingly a plan was devised for them to race up the English Channel in the hope – borne out, as it transpired – that the British would be unable to react with sufficient speed to launch any large-scale attack. The German plan involved many ships, including destroyers and E-boats to act as a screen, minesweepers to clear lanes, and anti-aircraft trawlers, and an air escort of no fewer than 282 Luftwaffe fighters. Astonishingly, this complicated and comprehensive plan was devised without a single known breach of security.

The British were of course aware of the possibility of the ships returning to Germany but thought this would be achieved by the long route (that is, northwards around the British Isles), although a contingency plan, designated

Operation Fuller, was prepared in case the shorter route should be attempted. Operation Fuller required the involvement of destroyers, motor-torpedo boats (MTBs) and large numbers of RAF aircraft, but it also included, almost as an afterthought, Esmonde's squadron of torpedo-armed Swordfish. For should the German ships come up the Channel, it was assumed they would do so at night, passing through the Straits of Dover two hours before dawn, and thus enabling the very vulnerable Swordfish to attack under cover of darkness, as they had done in their daring attack against the Italian fleet at Taranto the year before.

The initial Fuller plan required six Swordfish to attack in conjunction with a surface attack by destroyers and thirty-two MTBs, with air cover suppled by the RAF. As time passed, however, and the Germans failed to leave Brest, the planners gradually reduced the assets earmarked for the operation, withdrawing first the destroyers and then twenty-six of the MTBs. Despite this, the Swordfish remained on station and Esmonde continued to train his crews in night attacks.

The British had many intelligence assets monitoring the German ships – electronic surveillance of German radio frequencies, agents in and around Brest, submarines on patrol off the French coast, and various surface ships and aircraft on regular patrol. Despite all these precautions, however, the three ships sailed undetected at 2115 on Wednesday, 11 February 1942, initially proceeding due west and then swinging north-east into the Channel.

When it came, the British response was late, muddled and ineffective, and marked by a failure to realize what was happening, a lack of urgency, technical problems and sheer bad luck. By this time 825 Squadron had moved to Manston, just outside Ramsgate. On the morning of 12 February, Esmonde received five telephone messages. The first at about 1055, was from Admiral Ramsay's HQ at Dover, in which 825 Squadron was put on stand-by for a possible mission against unspecified enemy warships approaching the Straits of Dover from the west. At 1130 he received another call from Dover telling him that the targets had been identified as the *Scharnhorst* and the *Gneisenau*, and that they were proceeding at a speed of 21 knots. The third message came at about 1200 and was from 11 Group RAF, confirming that he would have three fighter squadrons as top cover and two fighter-bomber squadrons to attack the flak ships. Confident of such support, when he received the fourth call from a staff officer at Dover at 1215 with the query that 'The admiral wants to know how you feel about going in. It's your decision,' Esmonde replied, 'Yes.'

The die was cast. Esmonde briefed his squadron, speaking firmly and incisively, and telling them that they would attack in two sub-flights of three at a height of 50 feet. They would have fighter cover so they ought not to encounter any enemy fighters. The pilots were to select their own targets, with the proviso that they concentrate only on one of the two capital ships or the cruiser. In other words, they should not waste torpedoes on the smaller accompanying warships. The RAF station commander met Esmonde minutes later as he walked to his aircraft and was struck by his haggard appearance, ashen face and monosyllabic speech. But whatever it was that was troubling him, the naval pilot kept it to himself. It is not too fanciful, however, to guess that, having attacked *Bismarck* only a few months previously, Esmonde had a very good idea of what to expect. Then, just as he got into his cockpit, a runner arrived to tell him that yet another telephone message had been received, reporting that the speed of the enemy task group was 27 knots rather than the 21 knots advised. That meant that the ships were much further advanced than Esmonde had plotted, and time was now of the essence.

The six Swordfish took off at 1225, each carrying a single torpedo. Having assembled at a height of 1,500 feet, they then headed for Ramsgate where they circled to await the promised fighter escort. After ten minutes they were joined by just ten Spitfires of 72 Squadron, RAF, far fewer planes than had been promised. After several more circles, Esmonde made up his mind to go and, with a hand signal for the other aircraft to follow him, he turned towards the enemy and dived to 50 feet.

Esmonde's leading sub-flight was in line astern, with the second sub-flight behind in a V formation (known at the time as 'Vick'), and they flew on a bearing of 140°, the enemy being judged to be 23 miles away on that bearing.

The few Spitfires of the fighter escort were severely hampered throughout the operation, being heavily outnumbered by enemy fighters, having to cope with a very low cloud base (about 1,000 feet), and also having to try to match their speed of advance to that of the lumbering Swordfish. As a result, they were unable to prevent the first enemy attack which came about 10 miles out from Ramsgate, where Messerschmitt Bf-109s swept in from behind at sea level, hitting most of the Swordfish but inflicting no casualties and causing surprisingly little damage. They were quickly seen off by the Spitfires, but at this point (approximately 1250), with Esmonde roughly 2 miles from his targets, a new and much more numerous group of German fighters arrived on the scene and started to attack, although many of them had to lower their flaps and undercarriage in order to reduce their speed to something approaching that of their targets.

For the next eight minutes the six Swordfish plugged on towards their targets. These were now visible, and Esmonde altered course towards them, his two wingmen close to him in line astern as the German fighters approached for another attack. Esmonde's aircraft was the first to be seriously hit. A large section of the port wing was ripped off and then the tail caught fire. The gunner climbed on to the spine of the aircraft and, with a leg either side, clambered back to the tail, extinguished the fire and then returned to resume using his gun – an astonishingly brave act but also a telling comment on the slow speed of the Swordfish. The courageous gunner only survived a few seconds longer before he and Esmonde's observer were both killed outright, while Esmonde himself was hit in the back and the head. As he died, he managed to release his torpedo and then the plane crashed into the sea. The torpedo was aimed at the *Prinz Eugen*, but the German captain saw it coming and took successful avoiding action.

The two German fighters that had shot down Esmonde had also inflicted heavy damage on the two following aircraft. The second Swordfish, with its pilot wounded and its gunner dead, crossed the destroyer screen and headed towards the battle-cruisers, aiming for the *Gneisenau*. Just as the pilot released the torpedo, however, the aircraft was hit yet again, causing it to slew and the torpedo to head instead for the *Prinz Eugen*, which again took successful evasive action. The pilot managed to lift his aircraft over the destroyer screen on the far side of the German formation and then crashed into the sea, where he and his unwounded observer were rescued by a British MTB. The third aircraft also penetrated the destroyer screen and it, too, attacked and missed the *Prinz Eugen*, following which the pilot crashed into the sea, where all three aircrew were similarly rescued.

The second sub-flight saw the three aircraft ahead of them being torn apart, but although they manoeuvred violently to avoid enemy fighters and anti-aircraft guns, they did not falter in their advance towards the German ships. Having cleared the destroyer screen, however, they were then set upon by yet another group of fighters, which shot them down one after another before they could release their torpedoes. Thus, of the six aircraft that attacked, four were shot down and two were so badly damaged that they were forced to ditch. Of the eighteen aircrew, fourteen were killed and four survived, all from the first sub-flight. Two of the latter were badly wounded, but, astonishingly, two were completely unscathed. No damage whatsoever was inflicted on the German ships.

Esmonde was, without a doubt, both an excellent and a courageous commanding officer and a highly experienced and skilled aviator. By February 1942 he had been in command for eighteen months, during which period he had flown on operations in the Arctic, the Atlantic and the Mediterranean against a wide variety of targets, including *Bismarck*. He had also behaved with great coolness and courage during the final hours of the *Ark Royal*. Thus he had matured both as a commanding officer and as a man. Although still showing a great determination to 'get to grips' with the enemy, he was also respected and trusted by his men, and was known to outsiders as someone who ran a well-organized and efficient unit.

Esmonde was, however, a complex man. A citizen of the Irish Free State, he was under no obligation to fight a war in which his country was resolutely neutral. That he did so reflected not only a family tradition of fighting for the British Crown but also his belief that this particular war was a holy mission against a deadly evil – the figure of a crusader taken from his family crest and painted on his aircraft was a conscious choice.

Esmonde's superior during this operation was Admiral Sir Bertram Ramsay, a respected and admired leader. Even so, his message to Esmonde that 'the decision is yours' seems to have been fundamentally unfair. Ramsay is by no means the only senior officer to have placed such responsibility on a subordinate's shoulders, but senior officers are expected to make hard decisions, based upon their experience and knowledge, and delegating such a decision to a commanding officer who is due to take part in the proposed operation places an unnecessary additional load on him. For if the commanding officer says that he will not go, he is bound to be considered, by some at least, as overcautious or even cowardly, and if he is a professional officer then he may well consider that his future career depends upon the decision. Esmonde is on record as having once said that he was 'afraid to be afraid'.[4] It is therefore possible that, when faced with Ramsay's message, he was concerned that if he refused, it would be seen as a sign of personal cowardice.

It must be made clear, however, that Ramsay was a thoroughly decent and thoughtful man, and that at the time he sent the message, and, indeed, until well after the Swordfish had taken off, he had no reason to doubt that there would be five squadrons of escorting fighters, as promised. Indeed, the following day he commented: 'Had I known that the fighter escort might not keep their rendezvous I would have told Lieutenant-Commander Esmonde to remain on the ground; indeed, I would have forbidden the flight as an order.'[5]

What passed through Esmonde's mind during the operation will never be known. The plan had always been for a night attack. In the event, faced with a daylight attack on two capital ships and a cruiser, and having already flown against the *Bismarck*, Esmonde must have been fully aware that the odds were against him and his men, even if all five promised fighter squadrons did arrive. The RAF station commander at Manston was shocked to the core by Esmonde's appearance as he walked to his aircraft, and it would therefore appear that the naval officer had some idea of what he and his men were about to face. Despite this, he took off. A second opportunity to withdraw from the attack arose over Ramsgate, as he awaited the arrival of the escort. That he then continued was probably because of his driving sense of duty and the realization that his squadron now represented the only chance of stopping the German ships before they got through the straits and headed for Germany.

It should also be pointed out that the crews of the other five aircraft, equally aware that they faced overwhelming odds, never faltered and followed their leader without question.[6]

2ND BATTALION, THE DEVONSHIRE REGIMENT

Bois Les Buttes, 27 May 1918

The succession of commanding officers in the 2/Devons during the First World War has already been described in Chapter 3. Here the events surrounding the death of one of them are examined in more detail.

At the time that Lieutenant-Colonel Rupert Anderson-Morshead took command in April 1918 there was a general sense within the British Army in France that the fight should be carried on to the bitter end. On 12 April, the commander-in-chief, Field Marshal Sir Douglas Haig, issued an order that has since become famous: 'Every position must be held to the last man: there must be no retirement. With our backs to the wall, and believing in the justice of our cause, each one of us must fight on to the end.' This attitude worked its way down to the individual units. Indeed, it was taken so seriously that before one action in April 1918, the officers, sergeant-majors and platoon sergeants in the 1/Devons took a solemn oath to 'stand their ground unflinchingly at all costs'. At the time, Anderson-Morshead was still with the 1/Devons so he must have been aware of the oath.

On joining the 2/Devons, Anderson-Morshead was thrown straight into the battle at Villers-Bretonneaux (24–25 April), where for the first time in their history British infantry faced a tank attack. The new commanding officer and his men stood their ground, but at great cost, the battalion losing 5 officers and 142 men killed or missing, with a further 5 officers and 193 men wounded. These were typical of the losses that regularly occurred in infantry battalions at this stage of the war, and they meant not only the loss of experienced men and respected comrades, but also an influx of raw and inexperienced men who had to be absorbed and trained before the next battle.

By now the whole of the Eighth Division was in need of a rest and in the first week in May it was sent to a 'quiet' area in the French sector known as the Chemin des Dames. While en route on 3 May, the 2/Devons received a batch of reinforcements, a mixture of trained but inexperienced men from the Devonshire Yeomanry and a rather larger number of conscripts fresh out of training. The new divisional area had been quiet for some months, but it was on the direct route for any German attack on Paris, and the officers of the 2/Devons were given several lectures by French officers who emphasized its importance to France and who repeatedly stressed the necessity for the position to be held to 'the last man and the last round' should the German army attempt to break through.

The 2nd Battalion was part of the 23rd Infantry Brigade and on 20 May it went into brigade reserve, whereupon Anderson-Morshead immediately implemented a rigorous training programme, which included a battalion route march on 24 May led by the band and in which the men were smartly turned out, their brasses brightly polished. The training came to an abrupt halt two days later, however, when at about 1600 on 26 May, Anderson-Morshead received a message from brigade headquarters that a major German attack was expected at 0100 the following morning. He called the whole battalion together, briefed them as well as he could, and ordered a parade for 2000. After that what time remained was devoted to preparing for battle and to absorbing yet another large draft of reinforcements that had just arrived.

That night the brigade was deployed: the 2/West Yorkshires were placed forward in the 'outpost zone', with the 2/Middlesex in the 'battle zone' behind them, while the 2/Devons were held in 'brigade reserve', in positions around a low hill which, because of its shape, the Devons named Bois des Buttes (where this name came from is not clear, for it was certainly not the name used by the local villagers). This feature, the brigade commander told Anderson-Morshead, was to be defended 'to the last against all hostile attacks'. Unusually, the hill was honeycombed with a network of tunnels within which, when they arrived at about midnight, the 2/Devons took shelter in order to get some sleep. The only problem noted by the battalion was that, because they had moved up in darkness, only very limited reconnaissance of the tactical positions was possible.

Promptly at 0100, and precisely as 'intelligence' had predicted, a German artillery bombardment began which, as was usual at that stage of the war, combined gas and high explosive. The tunnels protected the 2/Devons from the high explosive but all had to don respirators against the gas. At 0345 Anderson-Morshead was instructed by brigade headquarters that the time had come for the battalion to move out of the tunnels and into the trenches. As they emerged, however, they encountered a dense ground mist that limited visibility and ground that was deeply cratered as a result of the artillery barrage. Worst of all, the Germans had already broken through the forward positions and were swarming towards the Bois des Buttes. As a result the forward companies were forced to take up defensive positions well short of those planned. Soon, small groups of men from the forward battalions appeared out of the murk and it became clear that all was not going well up ahead.

The fighting quickly broke down into a confusing series of conflicts between small groups and even, in some cases, individuals, and the greatly outnumbered Devons were gradually reduced in strength. Some men managed to get into trenches and resist as best they could, but most blundered about in groups, taking on similar groups of Germans whenever they bumped into each other in the mist.

Communications within the brigade were solely by runner, for there had been no time to lay cable, and a message from brigade headquarters received after the battle had begun ordered the battalion 'to stay put even if the Middlesex and West Yorks come back through you . . . you are not to retire'. No further messages were received, so Anderson-Morshead issued written orders that the battalion was to 'hold on at all costs'. However, as the fighting increased in severity and the Germans began to penetrate the position, the battalion was gradually split up. Some platoon commanders on the flanks, finding themselves cut off from the rest of the battalion, attempted to lead the few survivors to safety, but with mixed success. Eventually, the only organized resistance came from a group in the centre of the battalion position, led by the commanding officer and the adjutant.

It was here, at about 0930, that Anderson-Morshead saw a German horse artillery battery moving along a road below his position and led a group of sixty men in a charge, causing the German gunners to disperse in confusion. The Devons then returned to their position. Shortly after, a British artillery officer passed through and he later recorded what he had seen:

> They [the Devons] were entirely without hope of help but were fighting on grimly. The CO himself was calmly writing his notes with a perfect hail of high explosive falling all around him. I spoke to him and he told me that nothing could be done.

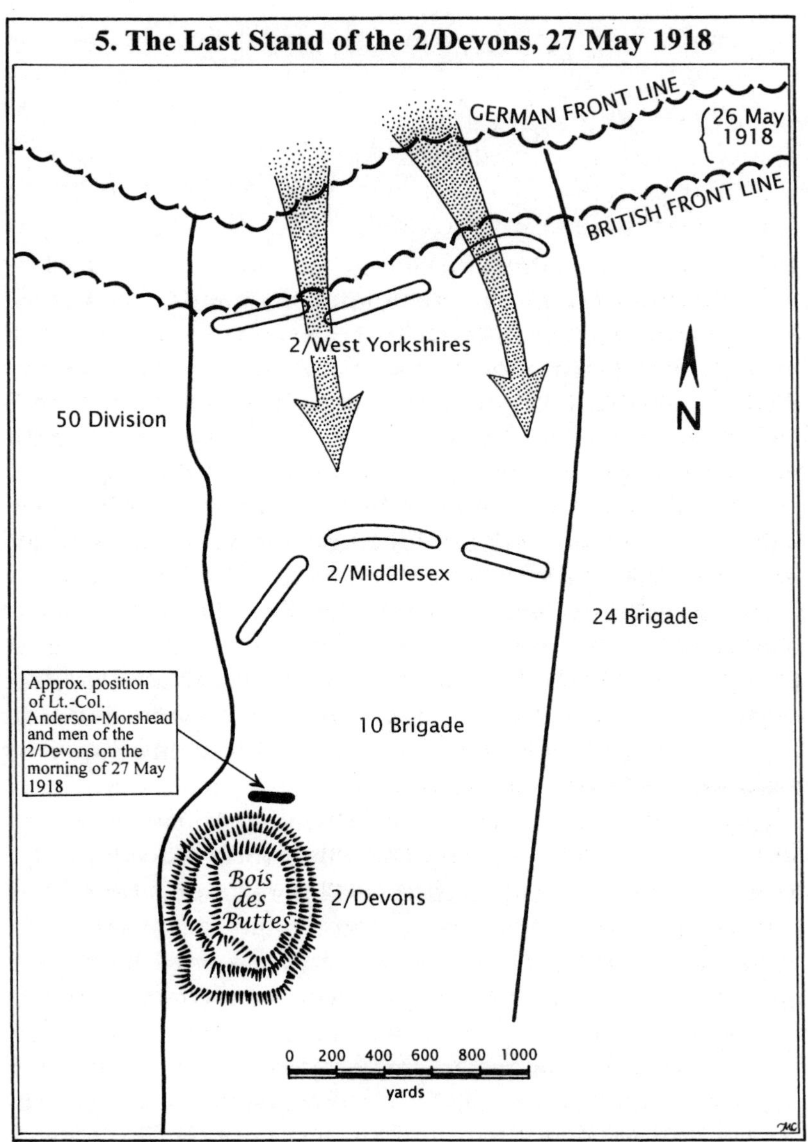

5. The Last Stand of the 2/Devons, 27 May 1918
GERMAN FRONT LINE
26 May 1918
BRITISH FRONT LINE
2/West Yorkshires
50 Division
N
2/Middlesex
24 Brigade
10 Brigade
Approx. position of Lt.-Col. Anderson-Morshead and men of the 2/Devons on the morning of 27 May 1918
Bois des Buttes
2/Devons
0 200 400 600 800 1000
yards

Then disaster struck. The adjutant, who survived the battle, described how

> The Colonel, seeing our men keeping their heads too low, got up on our parapet and directed our fire as successive targets appeared . . . About 11 a.m. our Colonel had gone round the trenches as several times before. Shortly afterwards when crossing the top of the hill, his usual route, he fell heavily forward and lay motionless.

The colonel was dead but the group, now commanded by the adjutant, fought on, its numbers steadily decreasing. Ammunition became so scarce that men had to strip rounds off the dead lying all around them, but the end could not long be delayed. Finally, at about 1230, with only a handful of rounds left, the remaining men fixed bayonets and charged down the hill towards the enemy, the few survivors of this last attack being rounded up and captured by the Germans. The battalion had been overwhelmed, but it had not surrendered. Between 0345 and 1230 that day the 2/Devons lost 28 officers and 552 soldiers, killed, wounded or missing. Those elements of the battalion that escaped regrouped, joined up with the men of the 2/Devons' rear echelon and carried on fighting for another eight days. The battalion then went into reserve but was quickly rebuilt, and on 26 July 1918 it was in action again, with a strength of 36 officers and 877 men.

Eighty-five years on, it is difficult to imagine just what it was like for commanding officers of the First World War who, aided by their officers, warrant officers and non-commissioned officers, held the battalions together throughout four years of war. Battalions like the 2/Devons were battle-hardened and experienced, certainly, but there was an incessant haemorrhaging of killed and wounded, which resulted in a constant need to absorb and educate an incoming stream of ill-trained and inexperienced though enthusiastic reinforcements. These had to be placed according to their abilities and the battalion's needs and, if all went well, there might be time for them to get to know their comrades in their platoons and vice versa. In the case of Bois des Buttes, however, there was scant opportunity, and the men of the latest draft were fighting for their lives within hours of arriving.

Why should commanding officers issue orders to 'fight to the last man' and, of equal importance, why do their men then do so? Doubtless, the general atmosphere of 'backs-to-the-wall', Haig's order and unofficial 'oath-taking' ceremonies such as that described in the 1/Devons played their part. In the case of the 2/Devons further pressure was brought to bear by the French and, finally, there were the brigade orders for this specific operation, charging the 2/Devons with defence of the position 'to the last'.

Nevertheless, the fact that they did so is remarkable, for the circumstances were hardly propitious. When the 2/Devons went into battle on the morning of 27 May, they did so at the end of a long and demoralizing retreat, during which the battalion had suffered heavy losses. There had been two large and very recent drafts of inexperienced men who diluted the pool of experience and who must have been bewildered by the events of 26/27 May. And to add to the battalion's difficulties, the officers had been unable to carry out proper reconnaissance of the planned positions. Despite all this, Anderson-Morshead and the 2/Devons stood and fought.[7]

COMMANDER LOFTUS JONES

The Battle of Jutland, 31 May 1916

At the Battle of Jutland on 31 May 1916 the British 4th Destroyer Flotilla consisted of HM ships *Acasta*, *Christopher*, *Ophelia* and *Shark*, the last commanded by the flotilla leader, Commander Loftus Jones, RN. The flotilla's task was to provide a screen force for the 3rd Battle Cruiser Squadron (3BCS) to protect it from attack by German cruisers and destroyers. At about 1730 Jones was steaming on the port side of 3BCS when he saw a formation of German warships in the distance (it was, in fact, the 2nd Scouting Group, consisting of four light cruisers and ten destroyers). Despite being outnumbered and outgunned, he immediately turned to attack, initially receiving fire support from 3BCS. *Shark* approached sufficiently close to launch a torpedo at the leading German cruiser but, as Jones turned his ship away, she came under heavy fire and was hit so severely that she was brought to a halt. The bridge, the engine room and the steering had all been damaged, while the forward gun was destroyed. As *Shark* lay helpless, another destroyer, *Canterbury*, deliberately exposed herself in order to lead the German light cruisers away, while *Acasta* approached with an offer of a tow. This was politely rejected by Jones, who told the captain 'not to get sunk for me'. *Acasta* was then almost immediately hit by two shells that caused extensive damage.

In a brief lull, Jones, already wounded, moved to the stern to continue conning the ship from the emergency steering position, but when some German destroyers approached and blew away the after gun, killing the crew in the process, he was forced to move to the midships gun to direct its fire. At this point Jones's leg was shot off at the knee, but a petty officer applied a tourniquet, enabling his captain to continue encouraging his men. Indeed, he even had one raise a new white ensign to replace the flag that had been shot away, thus ensuring that there was no possibility of the enemy thinking that his ship had surrendered. The midships gun continued to fire until *Shark* was hit by two torpedoes. Just before she rolled over and sank, a petty officer tied a lifebelt around Commander Jones, and as she went under he got him to a Carley float, where he survived for several hours before dying from exhaustion and loss of blood.

In carrying out a naval or military action 'to the last', the commanding officer undoubtedly makes a conscious decision, and in the examples given it is clear that the three commanding officers understood precisely what it was that they were asking their men to do. In all three cases, they also proved by their own example that it was a fate they were fully prepared to share with their men. Beyond that, the decision to continue to the end depended on a combination of leadership by more junior officers and the determination of the men themselves.

It should be noted, however, that, vital as the commanding officer may be, his death does not necessarily mark the end. Thus, while the crew of the *Shark* were inspired by Loftus Jones until their ship sank beneath them, the aircrews of 825 Squadron's second sub-flight continued to attack despite having seen their leader shot down in front of them. Similarly, the men of the 2/Devons continued to fight even after Anderson-Morshead had been killed.

Why do the men do it? In some cases it is clear that belief in a cause plays a part, as does national pride and naval or military discipline, but that cannot provide the whole explanation. The men of the 1/Foreign Legion, for example, discussed at the beginning of the chapter, had little interest in either some lofty ideal or the honour of their adopted nation, but they were proud of the traditions of the Legion, even though, at that point, it had existed for only thirty-two years. For the men of the 2/Devons their pride lay in a regiment that traced its existence back 233 years to 1685, a regiment that had fought at Dettingen in 1743, in the Peninsula between 1808 and 1814, and in many colonial campaigns throughout the nineteenth century. The aircrew of 825 Squadron belonged to a service whose entire tradition was based on attacking the enemy, and they were following a commanding officer whose leadership and airmanship they greatly admired.

However, even if it were possible to express all this in mathematical terms, these factors, in whatever proportions, take no account of two further elements. One, which is both unquantifiable and virtually indefinable, could be termed 'the human spirit'. The other, the vital spark that energizes all the rest, is the commanding officer, for without his orders and his example, nothing would happen.[8]

6

Bravery is not Enough

Gallant failures in war have a fascination all their own and many of them – and their commanders – are well known. Colonel Charlie Beckwith's attempt to rescue the American hostages held in the US Embassy in Tehran in 1980, for example, ended in a spectacular disaster at the temporary airfield known as 'Desert One'. The raid was dogged by a series of minor misfortunes, and when a helicopter crashed and burst into flames, cancellation was the only course left. Similarly, Colonel 'Bull' Symons raised a special unit to carry out a daring prisoner-of-war rescue operation at Son Tay in northern Vietnam in November 1970. The raiders reached the objective but the prisoners had been moved.

A lesser known but equally spectacular failure took place in 1941 when a special forces unit of the Royal Italian Navy attempted to penetrate the Grand Harbour in Malta.

COMMANDER VITTORIO MOCCAGATTA

Operation Malta Due, 26–27 July 1941

Capitano di Fregata (Commander) Vittorio Moccagatta commanded the Royal Italian Navy's special forces attack unit, Decima Flottiglia MAS (Tenth Anti-Submarine Boat Flotilla), usually known as Decima, which had its origins in the unit formed during the First World War to operate the 'human torpedo'. This device had been invented by the Italian Navy and was used, among other successes, to sink the Austro-Hungarian battleship *Viribus Unitis* (20,000 tons) on 31 October 1918. Interest in special forces operations was revived in the early 1930s and several plans were made for attacks on the British fleet in Malta during the 1935–6 Abyssinian crisis. This led to the formation of Decima.

By mid-1941 the unit was commanded by Moccagatta, who was neither a 'special forces' volunteer nor a man with any previous special operations experience. He had been appointed because he was an efficient and courageous officer, and an excellent organizer. Under Moccagatta's command Decima was split into two divisions, each commanded by dedicated and experienced specialists. The surface warfare division was led by Capitano di Corvetta (Lieutenant-Commander) Giobbe. This division had already carried out a successful raid against British shipping in Suda Bay in Crete on 26 March 1941, in which three small assault boats sank the cruiser HMS *York* (8,250 tons) and so severely damaged the tanker *Pericles* (8,324 tons) that it later sank under tow.

The underwater warfare division was commanded by Maggiore Genio Navale (Major of Naval Engineers) Tesei, a human torpedo specialist, who had already conducted several operations against Allied shipping in Gibraltar harbour. Tesei was dedicated to his beloved torpedoes, having played a significant role in their development, and he had spoken on several occasions of dying for his 'King and Country', particularly in an attack against Malta. Unknown to Moccagatta, Tesei had also been given a confidential warning by doctors that his previous activities had overstrained his heart and that if he took part in any more raids he might die.

Following the Suda Bay success, Moccagatta's next plan, devised in April/May 1941 but based on the 1935 project, was for an attack by assault boats on the British fleet in Malta, which was approved by the naval high command on 20 May, subject to surface reconnaissance, which was successfully carried out on 25 and 28 May. A subsequent aerial reconnaissance then showed that there were too few ships in Grand Harbour, leading to a postponement. The final go-ahead was not given until 23 June, authorizing an attack on either 27 or 28 June. The high command took the line that the proposed attack involved only a few men and resources, and that, if it were successful, it would have considerable benefits; while if it failed, losses in both lives and *matériel* would be small. Accordingly, Decima deployed to Augusta in Sicily, and carried out a further reconnaissance, led by Moccagatta, on the night of 26/27 June which got to within a mile of the harbour entrance and then returned. The Italians were not detected, but on the other hand they did not learn very much about the target either. The force sailed on 27 June, with motor torpedo boats towing the assault boats, but when one of the latter sank in rough seas the operation was called off. A second attempt on 29 June was also called off because of engine problems in several assault boats.

Moccagatta reconsidered his plan in the light of these experiences and made two major changes. The first was that, since the assault craft were clearly too small for an open-sea crossing even under tow, they would be carried by a 'mother ship' to a point as close to Malta as would be compatible with security. The second change was made at the insistence of the persuasive and fanatical Tesei, who succeeded in talking Moccagatta into agreeing that two human torpedoes would be included in the new plan and that one of them would be piloted by Tesei himself. This revised plan was approved for execution on the night of 25/26 July.

The entrance to Grand Harbour was flanked by the forts of Ricasoli and St Elmo, the gap between them being further narrowed by two breakwaters. In wartime this final gap was closed by a floating boom. The Ricasoli breakwater jutted straight out from the land, but the St Elmo breakwater was broken at its landward end by a narrow stretch of water which was intended to enable small boats sailing between Grand Harbour and the adjacent Marsamxett Creek to take a

6. Operation Malta Due: The Italian Plan

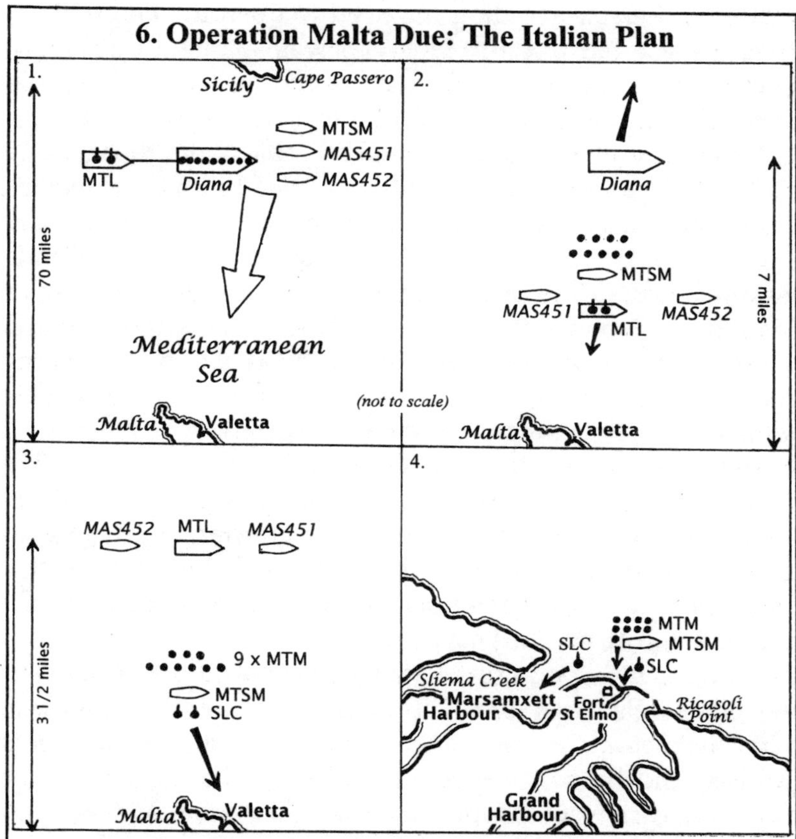

1. The flotilla sails after dark on 26 July. *Diana* carries nine MTMs and tows the MTL carrying two SLCs. The MTSM and two MASs proceed under their own steam.

2. Seven miles from the target, *Diana* offloads nine MTMs, casts off the MTL tow and returns to wait off Cape Passero. All the other boats proceed to the second waypoint.

3. At the second waypoint, 3 ½ miles from the target, the MTL offloads two SLCs. The two SLCs, the MTSM (Giobbe) and the nine MTMs proceed towards the target, leaving *MAS451* (Moccagatta), *MAS452* and the MTL to act as a rendezvous for survivors.

4. At the entrance to Grand Harbour, one SLC (Tesei) proceeds towards the St Elmo viaduct, while the second SLC makes a diversionary attack on the submarine base. Meanwhile, Giobbe in the MTSM waits a short distance away, ready to launch the MTMs as soon as the net is blown by Tesei.

short cut, thereby avoiding the need to open the boom every time one wanted to pass. This gap was spanned by a high-level viaduct, the centre of which rested on two large, cylindrical steel pillars, and was closed to unauthorized shipping by two heavy steel nets suspended by cables from the viaduct.

The Decima planners correctly identified this as the weakest link in the British defences and decided to breach the net under the outer of the two spans in order to obtain access for the assault boats to the targets inside the harbour. The assault boats and human torpedoes would be transported towards Malta, with the assault boats being offloaded at a point some distance from the harbour entrance, and the human torpedoes, which had a short range, taken closer in.

Also as part of the plan, the Italian airforce was scheduled to make three raids. The first was to be against Valletta at 0145, its main purpose to indicate the direction of the target to the naval force. It would be followed by two much heavier attacks, intended to divert the defenders' attention, on Valletta on 0230 and on Luqa airfield at 0430.

The force for Operation Malta Due (Malta Two) consisted of sixteen vessels (see Table 9) of different sizes and capabilities, with the actual attack being carried out by assault boats (MTMs) and human torpedoes (SLCs), while the remainder played a variety of supporting roles. The nine MTMs each carried a 660lb warhead and it was proposed that the pilot would approach the target at maximum speed. When about 110 yards away, he would lock the throttle, set the fuze, release the detachable, buoyant transom and then throw himself backwards into the water, where he would grasp the transom both to keep himself afloat and to provide protection against the shock from the explosion. The two SLCs were modified 21-inch torpedoes, fitted with seats and controls for a two-man crew. Their Italian nickname, *maiale* (pig), gives an idea of their handling characteristics. They had a very short range and on reaching the target the 550lb warhead would be removed and attached to the target's hull by clamps. The time-delay fuze would then be set and the crew would make good their escape.

TABLE 9: *Italian Vessels Involved in Operation Malta Due*

Vessels	Abbreviation	No.	Crew	Length	Speed (max)	Range	Role	Remarks
State yacht, *Diana*		1	250	374 feet	32 knots	Not known	Transport for MTMs	Built as state yacht/frigate
Motoscafi di turismo modificati (tourist motorboat, modified)	MTM	9	1	18 feet	33 knots	120 naut. miles	Assault on shipping in Grand Harbour	Built-in 660lb warhead
Motoscafi di turismo, silurante, modificato (tourist motorboat, torpedo-carrying, modified)	MTSM	1	2	23 feet	32 knots	200 naut. miles	Giobbe's command vessel	
Motoscafi anti-somergibili (motor torpedo boats, anti-submarine)	MAS	2	9	59 feet	42 knots	330 naut. miles at full speed	*MAS451:* Moccagatta's command vessel. *MAS452:* escort	Equivalent to British MTB
Siluri a lenta corsa (slow-running torpedo)	SLC	2	2	22 feet	5 knots	4 naut. miles at 5 knots, 15 naut. miles at 3 knots	*SLC1:* to blow hole in net. *SLC2:* diversionary attack	Detachable 550lb warhead
Motoscafo tourismo lento (motorboat, tourist, electrical)	MTL	1	6	31 feet	4-5 knots	Petrol 60 naut. miles; electrical 40 naut. miles	Carried Tesei and two SLC	One petrol engine/one electric motor

The largest ship involved was the *Diana* (2,550 tons), which had been built as Mussolini's state yacht but on this occasion was being used as an auxiliary vessel to carry the nine MTMs and their crews, and to tow the electric launch (MTL). The latter had been converted to mixed propulsion, with a petrol engine for the open sea and battery-powered electrical motors for a silent approach towards the target, although it was discovered that the petrol engine was not powerful enough and the MTL had to be towed most of the way, first by *Diana* and then by a torpedo boat.

The initial stages of the plan went surprisingly well. The force sailed from Augusta just after sunset on 25 July 1941, with Moccagatta aboard *MAS451*, the surface commander (Giobbe) aboard the MTSM, and the underwater commander (Tesei) aboard the MTL, which was also carrying his two precious SLCs and being towed by *Diana*. Morale was high and commitment in the unit so strong that the flotilla's medical officer had insisted on taking part and was aboard *MAS452*.

Despite the forecast, there was no moon, but the sea was calm. When the force arrived at a point seven miles from Fort St Elmo, the assault boats were offloaded from *Diana*. One, however, refused to start and so was immediately scuttled. *Diana* then set off towards Sicily to await the survivors' return, while the remainder proceeded to a second point, 3½ miles from Fort St Elmo, where the two SLCs were offloaded from the MTL. One of these was immediately found to have an engine defect, and even though its mission was only to carry out a diversionary attack, a full hour was devoted to efforts to repair the fault. The two SLCs eventually set off but the troublesome craft soon developed more faults and in the end the crew were forced to abandon it and make their way ashore, where they were eventually captured.

Aboard the other SLC Tesei steered towards Fort St Elmo and successfully reached the viaduct. Up to this point, the British, despite being on the alert, had not detected a single element of the attack force.[1] At 0447, twenty-two minutes later than scheduled, there was a huge explosion alongside the anti-torpedo net, in which both Tesei and his crewman died, almost certainly deliberately. Meanwhile, Giobbe in the MTSM had quietly led the remaining eight assault boats to within sight of the viaduct, still undetected by the defenders. When they saw the explosion set off by Tesei, the MTM pilots opened their throttles and launched their attack.

As he roared towards the viaduct, however, the leading MTM pilot realized that Tesei's warhead had failed to breach the net and that a second attempt was necessary, so he set the fuzes and threw himself clear just before impact. As no explosion ensued, it can only be assumed that the detonator failed. On seeing what had happened, Carabelli, the pilot of the second MTM, headed straight for the net and resolutely remained aboard his craft until it hit, apparently in a deliberate effort to ensure that the nets were breached. Unfortunately for this gallant officer, his attack precipitated disaster, for although the explosion succeeded in breaching the net, it was so violent that it brought the viaduct crashing down across the gap, though the structure may, of course, have been weakened by the explosion of Tesei's torpedo. Whatever the reason, the result was that the entrance to Grand Harbour was now completely resealed by a new and impregnable obstacle.

Meanwhile, the explosion of Tesei's warhead had alerted the garrison ashore. Grand Harbour now came alive. Searchlights swept the water, while the many guns and machine-guns blasted at any targets, real or imaginary, that were revealed. The remaining six MTMs, unable to get past the fallen viaduct, were caught milling around 500 yards off the shore and were destroyed within minutes. One pilot was killed and the remainder, all wounded, were soon captured.

Moccagatta had had a plan for recovering at least some of his men after the attack, although it had been recognized from the start that the MTM pilots would have no option but to become prisoners-of-war. This recovery plan was complex: the MTSMs were to collect the crew of one SLC while the MTL recovered Tesei and his crewman, plus the pilot of the leading MTM. These two boats were then supposed to rendezvous with the two MAS, where the MTL would be scuttled and the remainder would then return towards Sicily and a rendezvous with the waiting *Diana*.

In the event, no members of the attack force returned to any of the boats designated to collect them, but the boats duly rendezvoused with each other, the MTL was scuttled as planned and the remainder set off for Sicily. Unfortunately for them it was now light and, having been alerted by the events in Grand Harbour, British Hurricane fighters took off at dawn and swept out to sea. They quickly discovered the two MAS (one of them towing the MTSM) and in a series of attacks sank one MAS and seriously damaged the other. Twelve of the men in this group were killed, including both Moccagatta and Giobbe, but eleven survivors reached the MTSM, boarded it and, when the British had departed, headed north where they met the *Diana*. For some reason, she was never attacked by the British and, having picked up the survivors, she headed for home.

There is no doubt that every participant in Operation Malta Due showed great courage, that the unit was well-motivated and morale high. Equally, there is no doubt that it failed completely to achieve its aim of destroying British warships in Grand Harbour. Furthermore, the losses were heavy, and while the death or capture of the assault force had been expected, the deaths of the entire Decima command staff, including the commanding officer (Moccagatta), the commanders of both the surface warfare (Giobbe) and underwater (Tesei) divisions, and the doctor, had not. There were also heavy losses among those remaining offshore.

The basic concept of attacking the Royal Navy inside the strongly defended Grand Harbour was extremely daring, but that does not mean that it might not have succeeded, as Decima's later raid on Alexandria was to show.[2] There were, however, serious flaws in the plan. In particular, information about the enemy was woefully inadequate. The layout of Grand Harbour was well known from widely accessible maps

1. Lieutenant-Colonel Sir John Colborne, described by his contemporaries as 'the finest battalion commander in the Army', commanded the 52nd Foot at Waterloo. On his own initiative he carried out the daring manoeuvre that defeated the French Imperial Guard, swinging the battle in the Allies' favour

5. Wing-Commander Hubert Goodman commanded
103 Squadron for just five days in May 1944, losing his life on
his first mission on the night of 11/12 May, aged 29. He is seen
here with his wife Joyce, who had two children and was
expecting a third at the time of his death

6. Lancaster bomber PM-M² of 103 Squadron, the epitome of endurance in a long and hard-fought bombing campaign, in which units such as this suffered very heavy losses. Between May 1943 and December 1944, a variety of aircrew flew 140 operational missions aboard this aircraft, the highest total in Bomber Command

7. Lieutenant-Commander Baron von Forstner (*front row, centre*) with the crew of *U402* at its commissioning, 21 May 1941. These men carried out eight operational patrols in twenty-nine months, sinking 71,000 tons of Allied shipping. Their boat was lost with all hands on 13 September 1943

8. Commander Egan's ship in the battle against the Japanese submarine *I27*, HMS *Petard* had one of the finest records of any Second World War destroyer, being the only Allied warship to sink three enemy submarines, one each from the German, Italian and Japanese navies

9. and 10. Pictured from the British destroyer HMS *Petard* during the battle of 13 February 1944, the Japanese submarine *I27* runs on the surface, unable to dive. The submarine's 5.5-inch gun could easily have demolished the attacking British destroyers, but ranges were so close that every time a guncrew left the conning-tower they were killed by British gunners

11. Lieutenant-Commander Eugene Esmonde commanded 848 Squadron's attacks against the German battleship *Bismarck* (24 May 1941) and the battlecruisers *Scharnhorst* and *Gneisenau* (12 February 1942). A man of exceptional courage, he was awarded the DSO for the first of those attacks and a posthumous VC for the second

12. A Fairey Swordfish torpedo bomber, of the type flown by Esmonde in the attack on the German battlecruisers during the 'Channel Dash'. They were desperately slow and had to be flown at a height of 50 feet when launching the torpedo, making the three-man crew extremely vulnerable

13. In a painting by W. B. Wollen, Lieutenant-Colonel Rupert Anderson-Morshead commands the 2nd Battalion, Devonshire Regiment, at Bois des Buttes, 27 May 1918. Ordered to resist 'to the end', the CO and his men bought the time needed for other units to escape the German onslaught. The battalion was awarded a collective Croix de Guerre by the French for this epic stand

14. Commander Vittorio Moccagatta was a man of undoubted courage, who planned and led his unit during the Italian attack on the British naval base in Malta on 27 July 1941. But the plan was flawed and over-ambitious, and Moccagatta and many others paid with their lives

15. and 16. *MT16*, one of the tiny motor boats used for the Italian attack on Malta's Grand Harbour, shown shortly after her capture. It was intended that the pilot would sit in the stern and throw himself into the water just before the boat hit the target, when the strake running around the bow would trigger the high explosive with which the forward hull was filled

17. The P&O cargo-liner *Behar* was sunk in the Indian Ocean by the Japanese cruiser *Tone* on 9 March 1944. Of 104 survivors, 72 were murdered in cold blood aboard the *Tone* on the night of 18 March 1944

18. An officer who commanded several submarines during the Second World War, Captain Ariizumi Tetsonsuke of the Imperial Japanese Navy was a man for whom cruelty seems to have been second nature. He murdered most of his prisoners, and in August 1945 he himself committed suicide in order to avoid a war crimes trial

19. British sailors ashore on the fortress known as His Majesty's Sloop-of-War *Diamond Rock*, one of a series of water-colours by Johannes Epstein, who spent several weeks with the garrison. When the French managed to get ashore, the garrison's fate was sealed and its commander, Lieutenant Maurice, had no choice but to surrender

20. During the 1991 Gulf War, USS *Nicholas* (FFG-47) attacked Iraqi-occupied oil rigs, which led to accusations, subsequently dismissed, that the commanding officer had ordered his men to attack troops who had signified their wish to surrender

21. Captain Hans Langsdorff (*front row, behind blackboard*) with many of his crew aboard the pocket-battleship, *Admiral Graf Spee*. A greatly respected commanding officer, Langsdorff put the lives of his crew first and then took his own life to demonstrate his love for his ship

22. Lieutenant-Colonel Randal Austin (*right*) led his Marine battalion in a violent battle on Koh Tang Island on 15 May 1975. Rushed into action after very limited reconnaissance, having received an inaccurate intelligence briefing and obliged to deal with a convoluted upward chain of command, his task was far more difficult than it should have been

23. A situation not covered in Staff College training as Bob Stewart's progress is blocked by Croat refugees in Bosnia, 19 April 1993. Atop the Warrior personnel carrier are (*left to right*): Lance-Corporal Higginson, Dobrila Kolaba (interpreter, later murdered by persons unknown), Margaret Green (sitting, UNHCR), and Lieutenant-Colonel Bob Stewart

24. The minesweeper HMS *Fittleton* was attempting to transfer mail from HMS *Mermaid* on 20 September 1976, when she was swept under the bigger ship's bow and sunk. Twelve men died

25. The frigate HMS *Mermaid*. Note the high sides and the short fo'c'sle. The transfer station was just forward of the bridge

and charts, but up-to-date information on the layout and state of the defences, and on the actual shipping in port was not. According to Moccagatta's successor as commanding officer, Capitano Di Fregata Prince Borghese:

> The information we possessed regarding the state of the defences of the island was very sketchy: it consisted almost entirely of what could be deduced from air reconnaissance; it was incredible, but true, that we had no agent in Malta who could furnish us with intelligence! In particular, we were quite unaware of the new defensive expedients the British had put into operation, following the experiences they had undergone when Decima carried out the first attacks against Gibraltar and Suda Bay.

Moccagatta certainly undertook surface reconnaissance, but this achieved little more than a demonstration that vessels could get fairly close to Malta without any reaction from the British, while the aircraft sent to reconnoitre the harbour were chased off by RAF fighters. This meant, as Borghese pointed out, that among other factors Moccagatta was totally unaware that a new boom had been installed 300 yards inside the Fort St Elmo breakwater, which would have acted as yet another, and totally unexpected, obstacle to the MTMs.

That new barrier apart, the entire plan hinged upon one factor: the successful destruction of the barrier under the St Elmo viaduct. The slightest delay in achieving this, or a failure to do so, inevitably meant that the following craft would be concentrated in the killing zone of the British defences and thus come under immediate and very heavy fire. Next, the timings in the plan were extremely tight. The proposed time of the attack on the St Elmo nets was 0430. That was very close to first light and meant that the withdrawal of any survivors would have to be conducted in daylight and would therefore be vulnerable to air attack – as proved to be the case.

There has also been criticism, particularly in Italy, of the decision to incorporate the SLCs in the revised plan, which was made at Tesei's insistence. First, at a speed of 2.3 knots, Tesei's

SLC greatly delayed the arrival of the MTMs off the St Elmo breakwater; second, the delay caused by trying to repair the second SLC's faulty motor came at the worst possible time; and finally, the 'diversionary' attack, which may have looked significant in the planning stage, proved to be totally irrelevant in practice.

Finally, the personalities of the two key people in Decima, Tesei and Moccagatta, bear examination. Tesei had been warned by doctors that his body could no longer cope, and someone who knew him recalled that:

> He knew that he would no longer be able to dive, and indeed it took great efforts to convince the navy's leadership to let him participate in the mission. He had been planning and preparing it since 1935, and beyond it he had no future. He could not fail to die, since not to die would betray his true self and his ideals and, even worse, his SLCs. And to those who doubted the wisdom of the operation he replied: 'Whether or not we will actually manage to sink some ships, is not really important. What matters is that we must show that we are prepared to blow ourselves up in the face of the enemy.'[3]

A man with such a mindset is a danger on any operation.

Then there was Moccagatta. Before the operation, Giobbe, the commander of the surface force, developed serious doubts about the operation which he expressed privately to Moccagatta. The latter heard him out but then stuck by the plan and ordered Giobbe to carry on. Moccagatta was clearly a good officer and had the interests of his men close to his heart, as he demonstrated repeatedly during both the lead-up to the mission and its conduct. Nor was his courage in any doubt, as was proved by the reconnaissances and his presence during the raid. That does not mean, however, that he was the right officer for Decima, for though he was an excellent organizer his plan was overly complicated and contained a fatal flaw, while his control over his immediate subordinates, particularly Tesei, was inadequate.

Special forces units often complain that their superiors, who possess little understanding of their capabilities and limitations, interfere too much in the planning process. This was certainly not the case with Decima in which Moccagatta was given a very free hand. So far as is known, his superiors intervened only to the extent that they demanded two reconnaissances and insisted on a delay until sufficient targets were known to be in Grand Harbour. Indeed, it is clear that the plan would have benefited from a critical appraisal by a senior officer, who might have spotted its flaws.

Moccagatta should not have given in to Tesei's pleas to include the SLCs in the operation. These cumbersome machines added yet further difficulties to an already complicated plan, while their lack of speed, which was well known, coupled with the mechanical problems with one of them, caused fatal delays. In addition, the diversionary attack by the second machine proved to be totally unnecessary. The raid might still not have succeeded if the SLCs had been omitted, but that is no justification for their inclusion.

Moccagatta appears to have been unperturbed by the lack of intelligence on British defences, and his personal reconnaissance only proved that he could get close to the harbour entrance without being detected. It did nothing to demonstrate how the harbour defences were deployed or how alert the defenders were.

Finally, he failed to see that there are times when a leader must launch his men into battle and then stand back to await their return without becoming involved himself. His decision to remain on board *MAS451* off Malta is thus open to criticism, especially as he had no means of communicating with or influencing the men involved in the operation. To be fair, however, it is known that he had last-minute doubts and it may have been these, coupled with a desire to keep an eye on the over-eager Tesei and the reluctant Giobbe, that convinced him that he should remain with the raiders until the last possible moment.[4]

7

War Crimes

It is a principle of modern international law that those who violate the laws of war should be tried for their alleged crimes and, if found guilty, punished – a principle and an obligation which, it should be noted, apply equally to both sides. Such war crimes trials have several purposes, of which the first and most important is to establish the truth, since it is only by conducting an open trial, producing evidence and subjecting it to rigorous examination that the full extent of the crime can be understood by the general public. Thus, for example, at the Nuremberg Trials the full history of the gas chambers was set out in all its gruesome detail, not only for the sake of those present but also to establish a written record for the education of others around the world and for future generations.

War crimes can be committed by any member of the armed forces against members of the opposing armed forces as well as against civilians, as was shown in the recent conflict in the Balkans. Here, however, it is proposed to consider only war crimes in which the commanding officer was either the perpetrator of the crime or else was considered to bear an overwhelming responsibility for what happened.

Although the principle of trying war crimes goes back to the seventeenth century, it only became a reality at the end of the First World War when various Allied powers sought to bring a number of alleged war criminals to justice. After considerable discussion it was agreed that a state was responsible for trying its own nationals. As a result, strong pressure was brought to bear on the German government to bring to trial a number of men whose names were supplied by Belgium, France and the United Kingdom. The German authorities were very reluctant to act but eventually gave in and set up what became known as the 'Leipzig trials' in 1921, in which twelve men were accused of various crimes committed in the recent war. To the dismay of the Allies, the proceedings were farcical, for both the German state and the majority of its people were clearly averse to the whole process. The atmosphere may be judged from an incident in the case of a General Stenger, in which the public prosecutor's case rested upon the proposition that if the general was guilty he 'would have had the manliness to admit it' – a novel legal concept.[1] One of the other cases to reach trial was that of a U-boat commander and two of his officers.

Oberleutnant zur See Patzig

The Llandovery Castle *Affair*

On 27 June 1918 the German U-boat *U86* sank the British hospital ship *Llandovery Castle* with heavy loss of life and then killed many of the survivors who had taken to the lifeboats. The *Llandovery Castle* was a former liner that had been requisitioned early in the war by the British government and used as a troopship. Early in 1916 she was converted into a hospital ship to transport sick and wounded Canadian servicemen from Europe back to Canada. In accordance with the rules of the Tenth Hague Convention of 1907 the vessel was properly marked as a hospital ship and special lighting was fitted, while her name and description were passed through the proper channels to the enemy. The British also ensured that she never carried anything other than medical staff and sick or wounded servicemen.

In June 1918 *Llandovery Castle* made one of her routine

voyages, transporting Canadian sick and wounded to Canada, and then returned across the Atlantic without patients for her next assignment. On this voyage she carried the normal civilian ship's crew of 164, plus a medical staff of 80 Canadian Medical Corps doctors and 14 nurses, a total of 258. As was required by the rules for hospital ships, she sailed alone and did not zigzag. She had completed two-thirds of her passage when, on 27 June, in broad daylight, she was spotted by the German U-boat *U86*, commanded by Oberleutnant zur See Patzig.

There was no doubt aboard the U-boat that the vessel was a hospital ship, since her markings were perfectly clear, and her character was confirmed at dusk when her lights were switched on. Thus when Patzig announced his intention to attack, two of those on the bridge, Leutnant Dittmar and the coxswain Popitz, attempted to dissuade him. According to Popitz, Patzig knew that it was forbidden to attack hospital ships. However, the captain claimed that he suspected the ship was carrying not only munitions but also some United States airmen.

There had certainly been rumours within the Imperial German Navy and uncorroborated reports in German newspapers that Allied hospital ships were being improperly used to carry troops and munitions, but there had been no proof and the stories had never received any form of official endorsement. However, there had been dissatisfaction with Allied compliance with the regulations in one particular area, which had led the German Admiralty to declare that it would not comply with the rules concerning hospital ships in certain clearly delineated areas of the Mediterranean. The *Llandovery Castle* was, however, in the North Atlantic, and quite why Patzig believed she was carrying munitions was never established. Even more difficult to understand was his belief that there were American aviators aboard, since he had never seen the ship before and had received no radio messages to foster this belief.

Despite the advice of his subordinates, Patzig confirmed his decision to attack. At about 2130, 120 miles south-west of the Fastnet Rock, he submerged and launched two torpedoes, of which one hit *Llandovery Castle* amidships. The ship took about ten minutes to sink, during which time three lifeboats were launched. A further two lifeboats may also have been lowered, although what happened to them was never established – one theory was that they may have been dragged under as the ship sank.

The people in the three lifeboats that are known to have survived then set about rescuing some other survivors in the sea, but at this point Patzig brought *U86* to the surface and directed the lifeboat carrying *Llandovery Castle*'s master to come alongside his U-boat. Various officers were summoned aboard *U86*, where Patzig endeavoured to make them agree that both American servicemen and munitions had been aboard the hospital ship. The officers denied these allegations, however, and returned to their lifeboat. *U86* then cruised about the area for some time, stopping on several occasions to question yet more survivors. The occupants of the lifeboat commanded by *Llandovery Castle*'s master then heard repeated firing in the darkness, which could only have come from the U-boat, although since it was dark they were unable to see what was being fired at. Thirty-six hours later that lifeboat was rescued, but after a thorough search it was established that it was the only boat to have survived. Meanwhile, aboard *U86* Patzig made a false entry in the log and swore the three men who had been on deck with him to silence. Finally, he called his crew together and instructed them to 'forget' the events that had just taken place.

After the war the British were determined that the men responsible for this affair should be prosecuted. Patzig, as commanding officer, would be charged with ordering the sinking of a hospital ship, while the four who had been on the upper deck when the gun was fired (Patzig, two deck officers, Leutnants Dittmar and Boldt, and a petty officer, Meissner) would be jointly charged with the murder of the survivors in at least two lifeboats. Meissner, however, was dead and Patzig simply refused to attend, so only Dittmar and Boldt were actually arraigned on 21 May 1921 in front of the German Supreme Court at Leipzig. There were plenty of witnesses from among the crew of the U-boat and the hospital ship's survivors, but Boldt and Dittmar refused to give evidence on the grounds that they had given a binding promise of secrecy to their commanding officer.

After a full trial the court declared that 'the principal guilt lies with Patzig', who had clearly torpedoed a hospital ship, which ran contrary not only to international law but also to the specific orders of the German government and Admiralty. The two deck officers were found guilty of a lesser charge of aiding and abetting manslaughter (*Beihilfe zum Todschlag*) and were sentenced to four years' imprisonment and dismissal from the navy. Public reaction can be gauged by a German newspaper report headlined 'Four Years Imprisonment for U-Boat Heroes', and it is scarcely surprising that they were subsequently allowed to escape from prison after only a short time and that no effort was made to recapture them.

Patzig did not turn up for his trial, thereby abandoning his two subordinates, Boldt and Dittmar, to their fate. Their position was made even worse when, despite Patzig's behaviour, they adhered firmly to their promise to remain silent. Whatever else he had done, this was disgraceful behaviour on the part of their one-time commanding officer. Thus, in the absence of any other adequate explanation, it can only be concluded that Patzig knew perfectly well that what he did on the night of 27/28 June 1918 was wrong.[2]

KAPITÄNLEUTNANT HEINZ ECK

The Peleus *War Crimes Trial*

The events surrounding the 1921 Leipzig trials had a profound influence on the manner in which war criminals were tried in the aftermath of the Second World War. Rather than allow each nation to try its own war criminals, such trials were now the responsibility of the victorious powers. One of the first was convened by the British authorities in Germany in November 1945. Again, it involved a U-boat commander Kapitänleutnant Heinz Eck, commanding officer of *U852*.

Eck was commissioned into the German Navy in 1934 and then spent eight years in surface ships before transferring to U-boats in 1942. His general naval experience and seniority ensured that he went straight on to a commanding officer's course, which included a mandatory operational patrol aboard another U-boat as 'commanding officer-under-instruction', following which he was appointed to command *U852*. Eck's boat was a brand-new, long-range Type IXD$_2$ with a crew of fifty-seven officers and men, and his maiden voyage was to be to the German U-boat base at Penang, Malaya – no mean undertaking for a commanding officer on his first command.

The voyage started well. *U852* sailed from Kiel in mid-January 1944, successfully rounding the British Isles and reaching the Equator in early March without once being detected by Allied ships or aircraft. On the afternoon of 13 March 1944 Eck was on the surface heading south when the lookout spotted a tramp steamer proceeding on its own – a U-boat commander's

TABLE 10: *The* Peleus *Affair*

	U852	*SS Peleus*
Commander/Master	Kapitänleutnant Heinz Eck	Captain Minas Mavris
Completed	1944	1928
Displacement	1,616 tons	8,833 tons
Length	287 feet	Not known
Crew	57	35
Range, surfaced	32,300 naut. miles at economical speed	Not known
Weapons		
Guns	1 x 37 mm AA, 2 x twin 20 mm	None
Torpedo tubes	6 x 21-inch	

dream. Eck closed, unobserved, until at 1920, well after darkness had fallen, he launched two torpedoes at a range of 600 yards. Both torpedoes hit, and the steamer quickly sank without transmitting a U-boat sighting report. About twelve survivors (the precise number was never known for certain) were left in the water, clinging to rafts or bits of wood.

Eck then brought *U852* to the surface and cruised slowly among the debris and the survivors, instructing the engineer officer, Lenz, who could speak some English, to question the survivors. Lenz managed to establish that the ship was the Greek-owned *Peleus* under charter to the British Ministry of War Transport. Following assurances that Allied help would come the next day, the survivors were returned to their rafts. The U-boat then slipped away into the darkness. At that point, quite suddenly, and without consulting any of his officers, Eck ordered that machine-guns and hand-grenades be brought up from below. He then told the four officers standing with him on the bridge that the wreckage and the survivors were to be destroyed. Lenz, the engineer, who was on his seventh patrol and who was by far the most experienced U-boat man aboard, protested strongly that this was illegal, but he was overruled by his commanding officer. Dismayed, he went below.

The U-boat cruised slowly around the area, with the men on the bridge firing rifles and machine-guns and throwing hand-grenades at any wreckage they could see. They even considered using the 37mm cannon but were thwarted when they found that they could not depress it sufficiently to engage such close-range targets. After no less than five hours of this, Eck decided to leave the scene. As soon as they had submerged he spoke to the crew over the public address system, telling them that he had decided on the shooting with 'a heavy heart' but that it had been necessary. He then recalled recent losses of other U-boats in the area and ended by reminding them that their families back home in Germany were being pounded nightly by Allied bombers – war demanded hard decisions.

Eck had not, however, been as thorough as he imagined. Despite the five hours of shooting, four men were still alive.

One, the Third Officer, soon died, but the other three survived to be rescued by a passing steamer thirty-two days later. They duly filed statements concerning the events they had lived through, although none was able to identify the U-boat or name any of her officers.

Meanwhile, the British used Ultra to track *U852*'s regular radio reports, and when she was off the coat of Somaliland British aircraft from Aden attacked the boat, damaging her so badly that Eck had to beach her on Cape Gardefui, where his crew were captured. In the confusion Eck failed to destroy the ship's log which was found by the British. From this it was quickly established that *U852* was the submarine responsible for the massacre of the *Peleus* survivors. This was confirmed by the crew who, almost to a man, were highly critical of Eck and the other officers.

A full investigation was carried out but legal proceedings were not undertaken until after the war when a war crimes trial was convened in the British Zone of Germany in November 1945. Five men – Eck, the gunnery officer, the doctor, the engineer (Lenz) and one seaman – were charged with 'committing a war crime in that you in the Atlantic ocean on the night of 13/14th March 1944 when Captain and members of the crew of *Unterseeboot 852* which had sunk the steamship "*Peleus*" in violation of the laws and usages of war were concerned in the killing of members of the crew of the said steamship, Allied nationals, by firing and throwing grenades at them'.

The trial began with the defence drawing the court's attention to the wording of the charge, which appeared to indicate that it was the sinking of the *Peleus* that had been illegal. That would have meant that every submarine commander who had attacked a merchant ship would be guilty of a war crime. The court quickly agreed that the violations referred to concerned the act of firing at the survivors and not the original sinking of the ship, and on that basis the trial proceeded.

The prosecution produced sworn affidavits taken some months earlier from the three survivors, the only contentious issue being whether one account's inclusion of the dying words of the Third Officer was admissible or not. Having established that it was, individual evidence was given by five German sailers from *U852* who were then cross-examined by the defence counsel on minor points of detail.

The facts of the case were not challenged by the defence nor did Eck suggest that he had been following orders. He did not deny opening fire and using hand-grenades, and he agreed that survivors had been killed as a result of his orders. His defence rested on the grounds of 'operational necessity', for he claimed that he needed to destroy the rafts and other floating wreckage so that they would not give away the presence of his submarine. The gunnery officer and the doctor both claimed that they had obeyed the orders of their commanding officer, the doctor even maintaining that this overrode the regulation that forbade medical officers from bearing arms. The most curious feature of the case concerned the engineer, Lenz. He had argued with Eck about the legality of the order but had later returned to the bridge where he had fired a few rounds from the machine-gun.

The most telling piece of evidence against Eck was given by Korvetten-Kapitän Adalbert Schnee, a highly experienced U-boat officer who had commanded four U-boats and sunk twenty-four Allied vessels on sixteen patrols before joining Dönitz's staff in early 1944. Under cross-examination Schnee was asked what he would have done in Eck's position. After some hesitation he admitted that 'I would under all circumstances have tried my best to save lives, as that is a measure which was taken by all U-boat commanders. But, when I hear of this case, I can only think that Captain Eck, having been through a terrible experience, had lost his nerve.' This statement seriously damaged Eck's case, not least because he had not been through any unusually severe experience up to that point, nor had he been spotted by Allied ships or aircraft, while the attack on the *Peleus* had been straightforward. When pressed further by the prosecutor, Schnee concluded, 'I would not have done it.'

In his summing-up the Judge-Advocate drew attention to the fact that Eck had spent no fewer than five hours cruising around the area attempting to sink the wreckage. He also remarked that even if Eck had managed to dispose of the rafts and larger bits of wreckage, there would still have been smaller debris and a tell-tale slick of oil, which would have led an alert aircrew to realize that a U-boat had passed that way. Therefore, the Judge-Advocate suggested, a submarine commander who was genuinely concerned about the safety of his boat and crew would have left the scene at the highest possible speed. In other words, the defence of 'operational necessity' did not stand up.

The charge specified that the killings were 'in violation of the laws and usages of war', but there was no reference to any specific statute or legal code because none existed. It was therefore left to the court itself to take judicious notice of the precedents, to decide what the international 'laws and usages' were, and then to rule on whether the accused had been guilty of violating them. The Judge-Advocate did, however, make a telling observation in his summing-up when he stated that 'It is quite obvious that no sailor and no soldier can carry with him a library of International Law or have immediate access to a professor in that subject who can tell him whether or not a particular command is a lawful one.'

All five men were found guilty as charged and, after evidence in mitigation, Eck, the doctor and the gunnery officer were sentenced to execution by firing squad, while the other two were sentenced to imprisonment – the sailor to fifteen years and Lenz, the engineer officer, to a life term. The sentences were confirmed, and Eck and the other two officers were shot by a British firing squad in Hamburg on 30 November 1945. The two sent to prison, however, were released in December 1951 and August 1952 respectively.

Eck's behaviour after the sinking of the *Peleus* bears an uncanny resemblance to that of Patzig in the earlier war. Both cruised slowly among the survivors for a lengthy period in an attempt to eliminate the last of the survivors. Eck never challenged the prosecution's account of what had happened, his defence resting solely on what he claimed was operational necessity. The court judged otherwise.[3]

THE SUBMARINE PATROLS AND PRESUMED DEATH OF CAPTAIN ARIIZUMI TETSONSUKE

During his tour in command of the Imperial Japanese Navy (IJN) submarine *I8* Captain Ariizumi Tetsonsuke showed himself to be one of the most brutal commanding officers produced by any navy and by no means the most competent.

Ariizumi was born in 1904 and joined the IJN in 1920. He transferred to submarines in 1926, achieving his first command in 1938, but was then given a series of staff jobs from 1939 to early 1944, which must have been extremely frustrating for a man desperate to prove his worth in battle. Finally, however, he was appointed to command submarine *I8*. He sailed from Kure and arrived in Penang on 13 March 1944. From there he sailed again on 20 March on a mission to attack merchant shipping in the Indian Ocean.

Ariizumi's first victim was the Dutch *Tjisalak* (5,787 tons), with eighty-two aboard, which was hit by two torpedoes at

TABLE 11: *Captain Ariizumi's Submarines*

	I8	*I401*
Period in command	February–October 1944	February–August 1945
Completed	1938	1945
Displacement	3,583 tons	6,560 tons
Length	358 feet	400 feet
Crew	80	144
Range, surfaced	14,000 naut. miles at 16 knots	37,500 naut. miles at 14 knots
Weapons		
Guns	1 x twin 5.5-inch	1 x 5.5-inch
Torpedo tubes	6 x 21-inch	8 x 21-inch
Aircraft	1 x floatplane	3 x floatplanes

0545 on 26 March. The crew and passengers took to the lifeboats but an hour later *I8* appeared and all were ordered to board the submarine. There they were forced to hand over their lifebelts and all other possessions, especially jewellery, and were left only with the clothes they were wearing. Their hands were then tied behind their backs and they were ordered to kneel on the foredeck facing forward and told that if they looked around they would be shot. The abandoned lifeboats were pushed away and a few prisoners realized that the Japanese intended that they would not be required again. The Caucasians among the survivors were then pushed one at a time along the narrow catwalks to the after deck where they were either beheaded or clubbed to death, but when this proved too slow the remainder were tied together in groups and pushed overboard to drown. Astonishingly, five survived, were rescued and reported the barbarity. Many of the crew were so horrified by what had happened that Ariizumi found it necessary to make a broadcast over the loudspeaker to explain why he had issued such orders.

Ariizumi's second victim was the British *City of Adelaide* (6,589 tons), which was hit by two torpedoes at 1925 on 30 March after a 24-hour, 500-mile chase, sarcastically described by the submarine's engineer officer as the 'longest in history'. The ninety survivors abandoned ship in six lifeboats and a motorboat, and then watched as their ship began, very slowly, to sink. At 2030, *I8* appeared and shelled the still floating ship – and continued to do so for five hours. According to the survivors, the submarine crew must have seen them in the lifeboats but they made no attempt to interfere with them. When the merchant ship went down at 0230 *I8* abruptly submerged. The third victim of this patrol was a native dhow off Addu Atoll, after which Ariizumi returned to Penang.

Ariizumi sailed again on 9 June and attacked the British *Nellore* (6,942 tons) at 1815 on 29 June. Again, crew and passengers abandoned ship, but this, too, failed to sink so Ariizumi surfaced and shelled it until it went under at 2100. Two of *Nellore*'s boats later reached Diego Garcia with thirty-three survivors. There one of them, the chief officer, said that he had heard one group of survivors, which he knew included three women and two children, being summoned aboard the submarine. There is no official record of what happened to those who boarded *I8*, but in post-war interrogations several Japanese sailors claimed to have seen children being thrown into the sea.

Ariizumi torpedoed the American Olsen Line ship *Jean Nicolet* (7,716 tons) on 2 July. The ship sank slowly, enabling the survivors to reach the lifeboats and, as usual, he surfaced in order to finish off the burning vessel with gunfire. The survivors were then ordered aboard *I8* and the first, a young man aged 17, was immediately shot and his corpse thrown overboard. Having handed over their valuables the remaining ninety-five Americans had their hands tied and were made to kneel facing the bows. The master, three mates and the radio officer, who had identified themselves, were taken below, together with a passenger from the US State Department. The remainder were lined up and forced in turn to run the gauntlet of two rows of Japanese sailors armed with steel bars and wooden clubs. Those who fell were killed, one even being disembowelled with a sword, and their bodies thrown overboard.

The murder of those on deck was incomplete when aircraft engines were heard and the submarine's alarm bells rang. At this, the Japanese crew rushed down the hatches, leaving the remaining Americans on the foredeck, their hands bound behind them, and the submarine crash-dived. Despite the search for all their possessions, one resourceful man had managed to retain a knife with which he was able to free a few of his comrades. They then swam or clung to floating debris until a Royal Canadian Air Force flying-boat discovered them only a few hours later and dropped lifebelts and supplies. Summoned by the aircraft, HMS *Hoxa* arrived thirty-six hours later and rescued twenty-three survivors. On board the Japanese submarine, meanwhile, the prisoners who had been taken below were questioned and then all except the master and the official from the State Department were beheaded.

Thereafter, Ariizumi could find no more victims and he

returned first to Penang and then to Japan, arriving on 8 October. Somewhat surprisingly, both *Jean Nicolet*'s master, Francis O'Gara, and the State Department official survived the war and later recalled that *I8* had been 'quite filthy' and in a very bad state of repair, with pipes and other gear held together by wire.

In early 1945 Ariizumi was selected to command four large aircraft-carrying submarines, although still only with the rank of captain. His mission was to sail across the Pacific in order to attack a crucial lock on the Panama Canal, thus preventing US warships from transiting between the Atlantic and the Pacific. There were repeated delays and then the target was changed to aircraft-carriers in the US Navy's advanced fleet base at Ulithi, an atoll roughly half-way between Guam and Peletin in the Western Pacific. The submarines finally sailed in July with Ariizumi aboard *I401*. When *I401* reached the rendezvous, however, none of the other three submarines were there. Ariizumi then headed for the alternative rendezvous but he had still not reached it when, on 17 August, he received the radio broadcast announcing the end of the war, which was followed by orders to return home to Japan. Ariizumi proposed that the submarine should be scuttled with its entire crew aboard, but nothing had been decided when, on 27 August, *I401* met the American submarine *Segundo*, which ordered the Japanese boat to proceed in company to Yokosuka. A US Navy officer and six petty officers were placed aboard to ensure that this was done, and they remained on the bridge throughout the remaining four days of the voyage.[4]

On 30 August the Japanese crew were told that the US flag would be hoisted over their submarine at dawn the following day, before entering Yokosuka. At about 0300 Ariizumi went to the bridge, looked at the sea and the American petty officers, and then returned below to his cabin, where he shot himself. His body was found a few minutes later by members of the crew who wrapped it in a Japanese flag, took it through the after hatch and then threw it into the sea without the knowledge of the Americans.

Amongst Ariizumi's papers were found two letters:

To All My Superior Officers
I am extremely sorry to have to report that there were no War results obtained during this patrol and I feel the responsibility for this.

I die in accordance with Naval Tradition and to keep my pride as a Senior Commanding Officer in the Pacific Area up to the end of the war and in the belief that my picked subordinates will, as from now, be loyal subjects to the Emperor.

I pray for the re-establishment, in the future, of our Empire.

TENNO HEIKA BANZAI!

31 August　　　　　Naval Captain Ariizumi Tetsonosuke
(Signature)

To My Crew
Is there anything to be ashamed of in the actions of the Imperial Navy and the activities of this ship? I will purify it by spilling my human blood.

The source of a flourishing Nation is the superiority of its people.

Believe in the superiority of the people, co-operate and unite together.

Promote science.

Be tenacious and ferocious.

Naval Captain Ariizumi Tetsonosuke
(Signature)

Ariizumi's devotion to his Emperor and to his country's cause was absolute, but so was that of virtually the entire Japanese nation. Many Japanese officers were aggressive and many were also, at least by Western standards, cruel towards their captives, but Ariizumi stands out for what he himself described as his 'ferocity'. Indeed, his nickname among his peers in the IJN was 'The Gangster', doubtless based on Japanese perceptions of American gangsters as seen in pre-war

Hollywood movies and their predilection for shooting first and often, and asking questions only afterwards, if at all.

It is difficult to offer any explanation for Ariizumi's behaviour towards his captives, but there is no doubt that, as commanding officer, he was responsible for a series of acts of wanton cruelty. Even his crew were upset by the acts that they were ordered to commit, and Ariizumi had to address them over the submarine's loudspeaker after each killing to explain why such acts were necessary. Also, on each return to Penang he paraded the crew and swore them to secrecy before allowing them ashore.

Curiously, despite his professional studies and his insistence on the highest standards from his subordinates, Ariizumi was not particularly good at his job. In their post-war interrogations, his crew said that he seemed to be indecisive, and the engineer was particularly scathing in his description of the chase of the *City of Adelaide*.

After the war the Allies made strenuous efforts to trace the perpetrators of the many atrocities committed in the Indian Ocean. As a result six admirals, five captains, twenty-one other officers and eleven petty officers, a number of them from *I8*, were charged in a mass trial. Aspects of that trial made Allied investigators suspicious of Ariizumi's suicide. The case was, therefore, reinvestigated by a Japanese policeman who reported in September 1948 that he had found six witnesses from *I401*, all of whom confirmed that Ariizumi had killed himself. Whether the witnesses or even the policeman himself could be relied upon in such a matter is another question.[5]

Captain Mayazumi Haruo

The Behar Massacre, 18 March 1944

Captain Mayazumi Haruo of the IJN was commanding officer of the cruiser *Tone* when, in January 1944, she was sent to join Cruiser Squadron 16 (Rear Admiral Sakonju Naomasa) on a foray into the Indian Ocean, designated Operation *Sa-go* One. On arrived at the rendezvous Mayazumi was briefed by the admiral and his staff. His orders included a directive that merchant ships were to be captured, if possible, and brought back intact by prize crews to a Japanese-controlled port, but that only the 'minimum number of necessary prisoners' were to be taken for intelligence purposes, and that the remainder were to be 'disposed of'. Mayazumi expressed disquiet over the disposal of prisoners, suggesting that it would be more useful to bring them back and put them to work, but the admiral stated that these were orders from above and refused to discuss the matter further.

Three cruisers sailed on 28 February: *Aoba*, flying the flag of Admiral Sakonju, *Tone* (Captain Mayazumi), and *Chikuma*, sister ship to *Tone*. Once in the patrol area they steamed in a widely dispersed formation until, on the morning of 9 March, *Tone* sighted the British merchant ship *Behar*. As well as cargo she was carrying eighty-five crew, seventeen soldiers and sailors to man her defensive guns, and nine passengers (seven men and two women). *Tone* fired two warning shots, but when *Behar* tried to escape and transmitted an emergency signal, further shots were fired into the superstructure. Bowing to the inevitable, the master ordered that the ship be abandoned.

The survivors took to the boats but were intercepted by the Japanese and ordered aboard *Tone*. There they were greeted harshly, stripped of all but basic clothing, tied with ropes and forced to sit in rows on the quarterdeck, facing aft. Captain Mayazumi was convinced that his men were behaving with consideration, and eventually the prisoners were taken below to be packed into two small rooms. They were ordered to remain silent and any infringement of the rules resulted in an immediate beating.

There was no doubt among the Japanese that the order to 'dispose of' the prisoners meant that they should be killed, but Mayazumi's first signal to the flagship after taking the *Behar* simply informed the admiral that all 107 survivors were aboard *Tone* and that their interrogation was continuing. The reply was a curt reminder to retain the minimum number needed for intelligence purposes and to 'dispose of' the rest. Mayazumi ignored this very clear order, as he did a reminder received the following day, an act of disobedience which was against all his naval training and which he believed would lead to his immediate court martial on their return to harbour. Nevertheless, he was determined not to kill the prisoners if he could avoid it. In this he was supported by his second-in-command, Commander Mii, who reported to the captain that the matter had been discussed in the wardroom and that the majority of the junior officers were deeply opposed to the wholesale killing of defenceless prisoners.

Meanwhile, Sakonju decided to call off the operation and the squadron returned to Tanjong Priok roadstead on the island of Java, arriving on 15 March. The next day, a 'study conference' was held to discuss the recently concluded operation, at the end of which Mayazumi told the admiral that he did not wish to be responsible for executing his prisoners. The admiral gave ground slightly, agreeing that 'two or three' prisoners could be sent to his flagship for further interrogation. Mayazumi then persuaded a staff officer to increase the number to 'five or six', but on returning to his ship he actually dispatched fifteen. Curiously, despite the stated purpose of their transfer, they were never interrogated and after two days they were taken ashore to a naval barracks where they started a rigorous captivity that was to last until the Japanese capitulation in August 1945. Then *Tone*'s executive officer happened to meet Sakonju ashore and managed to obtain the admiral's agreement to send 'a few Indians, but no others' ashore. Mii rushed back to the ship and immediately dispatched seventeen before the order could be changed.

Mayazumi was now required to take *Tone* to the Japanese naval base at Shonan (Singapore) for a refit, but seventy-two prisoners were still aboard and he was given specific instructions by both the admiral and his staff to kill them during the voyage. *Tone* sailed on 18 March and Mayazumi selected Lieutenant Ishihara, the master-at-arms, for the deed, telling him that it was to be accomplished as humanely as possible and without firearms. Ishihara protested vigorously but was overruled by his captain. With considerable reluctance, he selected three other officers to assist him and on the fatal night the victims were summoned one by one from their prison below, as if for another interrogation, and taken to the quarter-deck. There each one was knocked out with a karate chop and his jugular severed with a sword. The body was then heaved overboard and the next victim summoned from below.

The crime came to light after the war when the fifteen surviving Europeans who had been sent ashore before *Tone* sailed were released from their prisoner-of-war camp and enquired what had happened to their shipmates. British and American intelligence experts carried out a search and eventually the Americans found the name *Behar* in *Tone*'s log and informed the British. Mayazumi was found to be working for the American forces in Japan and, when asked, did not hesitate to explain what had happened. As a result he and Sakonju were arrested and taken to Hong Kong where they were tried by the British at a 'minor war crime trial' that began on 19 September 1947.

Sakonju and Mayazumi, who were defended by Japanese lawyers, faced the standard form of indictment that they, 'in violation of the laws and usages of war, [were] together concerned in the killing of approximately sixty-five survivors from the sinking of the British motor vessel *Behar*'. The defence did not deny that the men (actually seventy-two) had been killed aboard the *Tone*, the arguments centering instead upon who had ordered whom to carry out instruction.

The hearing lasted for thirty days as the court sifted through the evidence with great care, but in the end both men were found guilty. Evidence of character was then given in mitigation of sentence and on 29 October 1947 the president announced the court's judgment. Rear Admiral Sakonju was sentenced to death and Captain Mayazumi to seven years' imprisonment. Following confirmation, Sakonju was hanged on 21 January 1948.

Sakonju received the death sentence because he declined to question the order, passed it on and insisted that it be carried out. Mayazumi's seven-year sentence can be taken as an indication that the court considered that his repeated efforts to prevent the order being put into effect went some way towards offsetting the fact that the killings eventually took place on board a ship under his command.

The original instruction that only the minimum number of prisoners be taken was based, at least in part, on the operational necessity of avoiding large crowds of prisoners on board warships which might subsequently find themselves in battle with Allied warships or aircraft. However, even if that premise had been valid at sea (and the question of 'operational necessity' was ruled illegal in other war crimes cases), it had certainly disappeared by the time *Tone* returned to Tanjong Priok. Indeed, Sakonju tacitly admitted as much by allowing thirty-two prisoners to be disembarked from the cruiser, fifteen to *Aoba* and seventeen directly ashore. Thus, when *Tone* went to sea again, the killing of the remaining seventy-two prisoners made no operational sense.

It is indisputable that Mayazumi put up a gallant and protracted resistance to an order he considered to be wrong. Nevertheless, while he was prepared to argue with Rear Admiral Sakonju, under whose command he was at the time, it does not seem to have occurred to him that he might raise the matter with Vice Admiral Takasu in Penang, although since he was the officer who had originally given the order, he was not likely to offer Mayazumi much help. Mayazumi might also have raised the matter with his usual commander but he was far away on the island of Truk. This point was put to Mayazumi by the president of the war crimes court, but the captain was scandalized: such a step, he assured the British lieutenant-colonel, would have been unthinkable in the Imperial Japanese Navy. Nor does it seem to have occurred to Mayazumi that, having sailed from Tanjong Priok with the seventy-two men still aboard, he could simply have taken them to Singapore and faced the authorities there with the problem. He would almost certainly have been court-martialled, but he would have stood by his principles. Instead, he decided that he had protested enough and that by his code he had done as much as he could for the prisoners. He had now been given a direct order which he could not disobey. It is likely that the war crimes trial judges agreed with him, for they gave him a very much lighter sentence than that awarded to his admiral.[6]

Quite where the dividing line can be drawn between legitimate military action and a war crime remains hard to judge, and it is clear that war crimes courts have themselves arrived at different conclusions even where the members of the different courts have been of the same nationality. Thus, one British court found Eck guilty and sentenced him to death for killing probably eight men, while another British court found Mayazumi guilty of killing nine times that number, but sentenced him to seven years' imprisonment. The deliberations of the latter court remain secret (and properly so), but it is clear that its members regarded Mayazumi's protracted disobedience in the face of Rear Admiral Sakonju's order as mitigating against his eventual compliance with what was clearly an illegal act.

It is significant that the crews of three of the submarines discussed told post-war interrogators that their commanders – Patzig, Eck and Ariizumi – had felt compelled by the obvious unhappiness among their crew to explain what had happened immediately after they submerged. Then, on return to port, all three had sworn their crews to secrecy.

Sinking enemy merchant shipping – whether by surface warship, aircraft or submarine – is accepted as a legitimate act of war. The problem arises with what happens afterwards. Eck, who was responsible for machine-gunning merchant navy survivors, was found guilty of a war crime, as Ariizumi would undoubtedly have been had he been caught. In an incident in the Pacific in January 1942, however, Lieutenant-Commander

'Mush' Morton, commanding officer of the submarine USS *Wahoo*, sank a Japanese troopship, a legitimate act of war, and then machine-gunned the many survivors struggling in the water. The men in the water were undoubtedly serving soldiers of the Japanese army, but there have been some suggestions that this was not a legitimate act of war. In any event, Morton died when his submarine was destroyed by Japanese aircraft on 11 October 1943, and the file was closed.

Most war crimes are quite clearly atrocities by any standard but there is a grey area. What may, to a commanding officer amid the stress and strain of battle, appear to be operationally necessary may well be judged quite otherwise in the calm, cool atmosphere of a war crimes tribunal. In such cases, commanding officers and their subordinates have neither the means nor the opportunity to consult the law or refer the matter to a higher authority for advice or a decision. Yet further complications arise when, as in Captain Mayazumi's case, it is precisely that higher authority which is ordering the crime to be committed. The extent to which an officer can be justified in disobeying an order is unclear, but it seems certain that at the *Behar* trial Mayazumi was considered far less blameworthy than his superior officer. This is a highly contentious area in which, even today, there are no clear-cut answers.

8

Surrender and Receiving Surrender

Commanding officers have to make many hard decisions, and one of the hardest is the decision to surrender. For the act of surrendering means that his unit is no longer able to perform its task. But, as will be seen, accepting surrender is not without its problems either.

On land, it is widely believed that a white flag indicates a wish to surrender. One of many such examples occurred at Beda Fomm in Cyrenaica on 6 February 1941, when British forces surrounded the retreating Italian army. One British officer who lay overlooking the scene later recalled:

> I watched with apprehension the movements of the scene before me . . . Our threat was but a façade – behind us there were no more reserves of further troops . . . Gradually I became aware of a startling change. First one and then another white flag appeared in the host of vehicles, until the whole column was a forest of waving white banners.[1]

This concept is, however, only partly correct. By the custom of land warfare the white flag actually indicates a wish to talk (to 'parley' in the old-fashioned usage), which may be for a variety of reasons, of which capitulation is but one. Others include a temporary ceasefire to enable the wounded to be collected or the dead to be buried, or to allow non-combatants to leave a besieged garrison. One example of the use of the white

142

flag in this way occurred on 22 December 1944 during the Battle of Bastogne, when the attacking Germans sent a *parlia-mentaire* consisting of two officers and two soldiers who carried a written message from the German commander to the enemy. The party crossed the lines under the white flag which, despite the bitterness of the fighting, was scrupulously respected. The four Germans were held for several hours with complete propriety while the message was delivered to the divisional commander, and when the written reply was received they were returned safely to their own lines. (The German note demanded the surrender of the Americans to which the formal reply was the single word 'Nuts!', but that is another story.)

Apart from the white flag, indications that individuals intend to surrender take various ad hoc forms, including standing up without weapons or raising arms in the air. On several occasions during the Iran-Iraq War of 1980–8, Iraqi units simulated surrender as a *ruse de guerre* and then, having lulled their enemy into a false sense of security, opened fire and attacked, a flagrant breach of international law. Then, in the very early stages of the Gulf War in 1991, the Iraqis came up with a novel ruse when a battalion of tanks drove towards Saudi Arabian positions with their turrets reversed and their guns pointing to the rear. This caused considerable confusion among the Saudis who believed that the Iraqis were surrender-ing and so withheld their fire. The tanks were thus able to get close to the Saudi positions without harm, whereupon they rotated their guns forward and went into the attack. No white flag had been displayed and it is debatable whether an advanc-ing tank, with its gun reversed, should properly be understood to be surrendering, but the stratagem certainly gained the Iraqis a valuable few minutes.

Perhaps the clearest sign that an enemy wishes to surrender occurs at sea where the hauling down of a ship's colours is a universally accepted signal that the senior surviving officer wishes to end further resistance. Some navies, including the Royal Navy, are so apprehensive that a single flag might be shot away and thus give a mistaken impression of a readiness to surrender, that they fly three or more 'battle ensigns' when in combat. Other signals of naval surrender are for a submarine to surface, or for the crew to take to the boats. It is, of course, impossible for an individual or a small group to surrender at sea. Either the senior officer surrenders his ship in one of the ways described, or the fight goes on.

There is no universally recognized means of surrendering in the air, since it is almost impossible for an aircraft to indicate that it has ceased to be able to fight. As a consequence, it is generally considered acceptable that attacks against a damaged aircraft should be pursued either until it has crashed or until a crash has become inevitable, although in no case is it acceptable to fire at the crew once they have baled out.

Nor is the question of accepting surrender a straightforward issue. Clearly, a commanding officer must not prejudice the success of his own tactical operations, nor should he place his own men's lives in peril. However, according to international law, once an individual or group has shown a clear intention to surrender without conditions, the commanding officer is obliged to cease firing and to afford them the protection guaranteed to prisoners-of-war.

Surrender is therefore an area that is, at best, fraught with difficulty and which can, at worst, lead to misunderstanding or disgrace, and it is certainly one with which no commanding officer ever wishes to be faced. Two examples are given here. In the first it is clear that the commanding officer and his men had done all that could be expected of them. The second example is, however, much more controversial.

In accepting surrender, the commanding officer has to make a quick decision, balancing the safety of his own men against the rights of his opponents. The issue is by no means always a simple one, as an example taken from the Gulf War will illustrate.

SURRENDERING WITH HONOUR

Lieutenant Maurice and HMS Diamond Rock, *1803–4*

In late 1803 Commodore Sir Samuel Hood, aboard his flagship HMS *Centaur*, was blockading the French harbours of Port Royal and St Pierre on the island of Martinique, when one of his ships visited a deserted island named Diamond Rock to look for anti-scorbutic plants to prevent scurvy among the crew. The island lay just under a mile off the south-west tip of Martinique. On returning to the squadron the ship's captain briefed the commodore, and a plan was conceived to place a garrison on the rock, whose task would be to dominate the channel between the rock and Martinique. Lieutenant Maurice was appointed garrison commander and the follow-ing week he took a working party ashore. There they first arranged accommodation and then, since there was no natural source of water, constructed a large tank to catch the overnight dew. Next a heavy gun was mounted at sea level (Centaur Battery) and a 32-pounder carronade was raised on a jack-stay to another rock 360 feet above sea level (Hood's Battery).

This, however, was not sufficient, for Hood's plan involved mounting two guns on the summit, which was achieved with a typical combination of naval ingenuity and engineering skill. *Centaur* was moored close to the precipitous southern face of the rock and the shore party dropped a light line from the summit to the ship, which was used to haul aloft a heavy cable. This was made fast to a projecting rock 80 feet below the summit and then served as a jack-stay, from which a traveller was suspended. A 5-inch hawser was then hauled up to the summit, riven through a pulley and both ends returned to the ship, where one was secured to the traveller and the other passed around the capstan. Next each 24-pounder cannon was suspended in turn from the traveller and a work party started to turn the capstan which pulled the gun slowly upwards. Each gun had to be raised 700 feet vertically and 400 feet horizontally, which took all day, and then, once at the top, it had to be man-handled to the summit. In all it took two days to mount both the guns, following which, with a touch of whimsy, the commodore commissioned the garrison 'His Majesty's Sloop-of-War *Diamond Rock*' with Maurice as its commanding officer. The sailors were happy to be ashore for a change and worked with a will. One of the caves was converted into a hospital, some bullocks and sheep were landed to graze, and numerous visitors called to see this unusual 'warship'.

HMS *Diamond Rock* was used as an aggressive base. In its first operation a raiding party was sent ashore on Martinique to capture a group of French engineers who were preparing to recapture the rock. Next, on 3 February, four boats were rowed 20 miles to Port Royal Harbour where the crews captured the French corvette *Curieux*, which was pressed into British service. Over the ensuing months the British garrison became a thorn in the side of the French and severely interfered with their use of Martinique's harbours, just as had been planned.

Word of this British impudence eventually reached Napoleon himself who determined to act. A squadron was assembled at Port Royal on Martinique, consisting of five warships and a landing force of several hundred soldiers. The squadron sailed on 29 May and arrived off Diamond Rock on 31 May. Maurice, seeing eleven gunboats full of troops heading for the rock, evacuated the lower level, raised the ladder and awaited developments. The attackers landed with some difficulty at about 0900 under heavy fire from the British on the heights. Once ashore the French commander Boyer soon discovered that what had appeared from a distance to be an easy task was in practice much more difficult. 'The moment we landed,' he reported later, 'this illusion ceased – I saw nothing but immense precipices, perpendicular rocks, a threatening enemy, whom it was impossible to reach, and insurmountable difficulties on all sides.'

The 120 men of the British garrison maintained a hail of fire with muskets, cannon, rocks and caskets filled with stones, and, to add to the attackers' discomfiture, they compelled the boats to put out to sea without having landed any provisions.

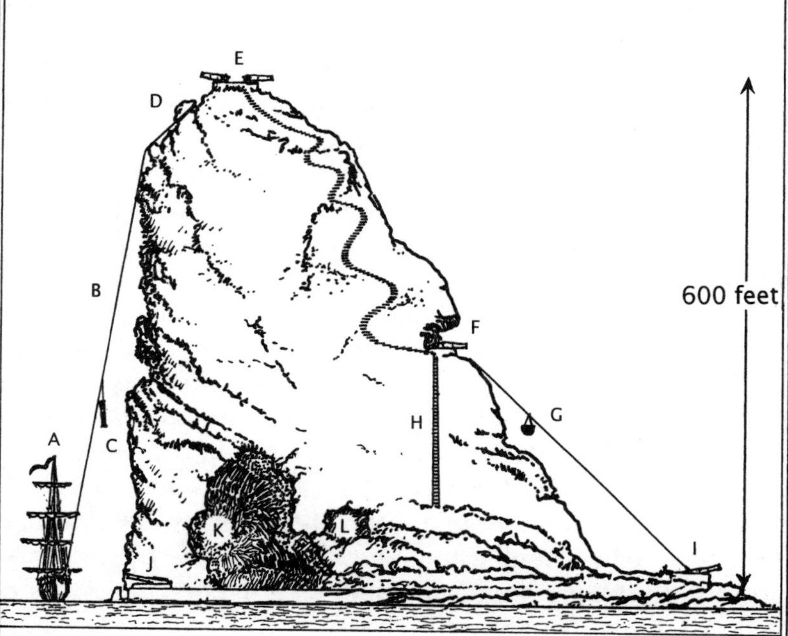

7. HMS *Diamond Rock*, 1803–4

A. HMS *Centaur*
B. Jack-stay
C. 24-pounder gun ascending
D. Upper end of jack-stay secured to projecting rock
E. Upper battery (two long 24-pounders)
F. Hood's Battery (24-pounder carronade, plus ammunition and stores)
G. Jack-stay with large tub for lifting stores to Hood's Battery
H. Jacob's ladder (raised every night)
I. Centaur Battery
J. Queen's Battery
K. Hospital Cave
L. Sleeping Cave

Boyer withdrew his men into the two large caves that had been used as a hospital and as sleeping accommodation, although an enterprising lieutenant managed to ascend the rocks, accompanied by twenty-five men. Unfortunately, they were discovered by Maurice's men and captured, but they managed to escape a few minutes later and returned to the lower level, having lost two men killed and two wounded. In the caves the French soldiers expressed considerable dissatisfaction; they had been seasick for two days on the voyage and had eaten nothing; now they were ashore they still had nothing to eat. In addition, there were many wounded among them and there was no water except for a few barrels of putrid liquid abandoned by the British.

During the night, however, a French boat managed to reach the shore, bringing six grenadiers and some desperately needed rations, and on the following night more grenadiers managed to get ashore. The standoff continued throughout the following day, but then a route was found up the rocks and a party managed to establish themselves close to Hood's Battery.

The British were now in a difficult situation. First, a round from one of the bombarding ships had hit the upper water tank, splitting the concrete and allowing the precious stock of water to drain away. Second, the expenditure of ammunition and powder had been far greater than anticipated and reserves were low. It was also clear that, despite their casualties, the French were firmly lodged ashore and defeat was inevitable. Consequently, Maurice decided that his men had done enough to satisfy the demands of honour and so, at 1630, he 'threw out a flag of truce'. As was the custom of the day, Maurice drafted his own terms, which included surrender with honour, the security of private property and transport (at French expense) to the British island of Barbados, to all of which the French commander was happy to agree.

Once in Barbados, and again according to custom, Maurice and his ship's company were court-martialled in order to apportion any blame. The court duly sat on 24 June 1805 and, having heard the evidence, concluded that the Officers and Company of His Majesty's late sloop *Diamond Rock* did everything in their power to the very last, in the defence of the Rock, and against a most superior force, and Captain Maurice behaved with firm and determined resolution and did not surrender the *Diamond* until he was unable to make further defence for want of water and ammunition; the Court do therefore honourably acquit Captain Maurice accordingly.

Maurice's defence and surrender followed the custom of the time: a besieged garrison or warship would put up the stoutest defence possible, but when circumstances dictated that defeat had become inevitable, then it was both permissible and honourable to surrender. Indeed, it was the unwritten but widely observed custom on land that if a besieged garrison (for example, a fortress) negotiated a surrender at an early stage, then the safety of the troops and the civilian populace were guaranteed. If, however, the besieged garrison resisted to the very end, then the attackers were justified in putting any surviving troops and civilians to the sword and sacking the place.[2]

A TRAGIC ERROR

The 'Two Colonels' at St Quentin, 27 August 1914

The affair of the 'two colonels' at St Quentin in August 1914 was quite a different matter. Great Britain declared war on 4 August 1914 and immediately dispatched the British Expeditionary Force (BEF) to France where the forward divisions were rushed by train from the Channel ports to take up their positions in the front line. The BEF acquitted itself well in the first major battle of the war at Mons (23 August) but when the French Fifth Army on its right flank withdrew without warning, the BEF were forced to follow suit. By 25 August the British II Corps, commanded by Lieutenant-General Horace

Smith-Dorrien and consisting of the 3rd and 5th Divisions, was withdrawing through Le Cateau.

That afternoon Smith-Dorrien was ordered to continue the withdrawal, but as his troops had fought a major battle and then withdrawn forty miles on foot in four days, all with virtually no sleep, he decided to make a stand at Le Cateau in order to gain a brief respite for his exhausted men. The Battle of Le Cateau on 26 August lasted all day and involved both II Corps and three other formations which had until then been under the direct command of BEF HQ but which were now placed under Smith-Dorrien's command. These were the Cavalry Division and the 19th Independent Brigade, which were already deployed, and the 4th Division which had just reached Le Cateau by train from the Channel ports. The battle involved a number of fierce hand-to-hand engagements but was predominantly an artillery engagement over open, rolling downs with few woods and little dead ground to give cover to infantry. The Germans employed over 300 guns and 8,000 British troops were lost, 40 per cent of them to artillery fire. Having halted the full strength of General Alexander von Kluck's First Army and forced it to deploy and fight, Smith-Dorrien continued the retreat on the following day.

One of the many incidents that occurred during these momentous events involved two British infantry commanding officers: Lieutenant-Colonels John Elkington of the 1st Battalion, Royal Warwickshire Regiment (1/RWarwicks), and Arthur Mainwaring of the 2nd Battalion, Royal Dublin Fusiliers (2/RDublin). The two men had much in common. Both belonged to families with strong military connections and both commanded regiments which their fathers had commanded thirty years earlier. Both were also graduates of the Royal Military College, Sandhurst. Mainwaring was commissioned in 1885 and Elkington in 1886, following which they served in various parts of the British Empire until achieving command, Mainwaring in 1912 and Elkington in February 1914, although the latter was the senior, having been promoted to lieutenant-colonel in 1910. Thus, they were very experienced officers, thoroughly steeped in British military traditions, and each had achieved the most cherished dream of all British regimental officers of the day – command of their battalions. They were however somewhat elderly for such active posts. Elkington was 48 years old. Mainwaring was 50 and also suffered from colitis.

The units to which each belonged were regular battalions, part of the 10th Brigade (Brigadier-General J.A.L. Haldane) which was in turn part of the 4th Division (Major-General T. d'Oyley Snow). In peacetime, the brigade had been stationed in Kent, but shortly after the outbreak of war it was sent to north-east England on anti-invasion duties, with the 1/RWarwicks and 2/RDublin sharing Strensall Camp outside York. A week later, however, the brigade was sent south again to go to France.

The 10th Brigade disembarked at Boulogne on the evening of 22 August and marched to a reception camp outside the town, arriving there in the early hours of 23 August. This was the day of the Battle of Mons, and the new arrivals were still getting themselves sorted out when, at midday, the brigade was ordered to move immediately to the front. The battalions entrained that afternoon, spent the night on the trains and arrived at Le Cateau at about 1000 on 24 August where they detrained. The town was crowded with the first troops retreating from Mons and conditions were chaotic, but the 10th Brigade marched in good order a few miles north-westwards to bivouac in the area of Beaumont.

Late that same evening the brigade received orders to move forward and act as rearguard to II Corps, whose 3rd and 5th Divisions were withdrawing amid scenes of some confusion. So, at 0200 on 25 August, the brigade marched north through Viesly and at 0600 halted for breakfast near St Python. They were still there at 0700 when gunfire was heard to the north, and at 0800 the brigade commander ordered a short withdrawal to a new defensive position where they dug in. They were still digging when some retreating British and French troops passed through, but although the 10th Brigade came under heavy artillery fire, they saw no sign of major bodies of either friendly or enemy units.

At this point the three brigades of the 4th Division were deployed one in front of the other, with the 12th Brigade at the rear around Viesly, the 11th Brigade in the middle and the 10th Brigade to the fore, just south of St Python, although the 19th Independent Brigade was even further forward in the area of Haussy. The 19th Brigade then withdrew through the 4th Division and Major-General Snow ordered his troops to follow suit, starting at 2100 that evening. The 12th Brigade led the way, followed by the 11th Brigade and finally the 10th Brigade serving as rearguard, the last's route taking it first to Beauvois and thence to Haucourt, which was reached at about 0500 on 26 August.

This was the day of the Battle of Le Cateau, in which the 4th Division formed the left flank of the British position, the engagement taking place in torrential rain that turned the fields into a quagmire. Mainwaring's first action on arriving in Haucourt was to send his adjutant to brigade HQ for orders, but it could not be found, although a divisional staff officer did visit battalion headquarters to ask Mainwaring whether his men would be ready to march another 12 miles, starting at 0700. Mainwaring replied that provided his men could have a short rest and something to eat there would be no problem, but at this point (around 0600) a firefight started about 1,000 yards to the north-east which caused the 2/RDublin to take up defensive positions and deprived the men of the opportunity to have breakfast. All of the 10th Brigade was then shelled for several hours, during which numbers of French troops came pouring over the hill to the front.

During the morning Mainwaring received two messages, both from divisional staff officers. The first, from the divisional commander's ADC, was that 'the general wishes you to hold on here to the end'; the second, from a staff officer, read 'It's going to be a case of long bowls; no retirement.' By this stage the men were very tired and were falling asleep despite the shelling.

At 1700 two artillery batteries which had been in front of the

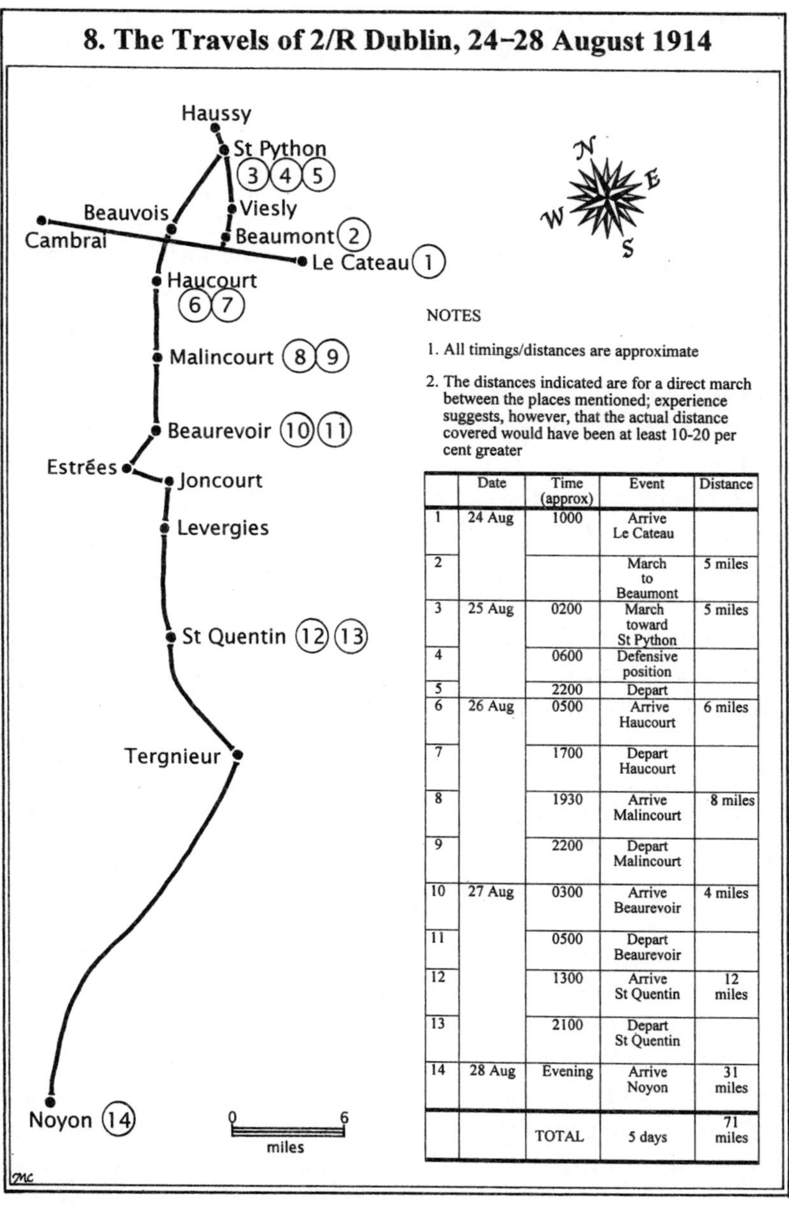

	Date	Time (approx)	Event	Distance
1	24 Aug	1000	Arrive Le Cateau	
2			March to Beaumont	5 miles
3	25 Aug	0200	March toward St Python	5 miles
4		0600	Defensive position	
5		2200	Depart	
6	26 Aug	0500	Arrive Haucourt	6 miles
7		1700	Depart Haucourt	
8		1930	Arrive Malincourt	8 miles
9		2200	Depart Malincourt	
10	27 Aug	0300	Arrive Beaurevoir	4 miles
11		0500	Depart Beaurevoir	
12		1300	Arrive St Quentin	12 miles
13		2100	Depart St Quentin	
14	28 Aug	Evening	Arrive Noyon	31 miles
		TOTAL	5 days	71 miles

2/RDublin all day withdrew, leaving two disabled guns behind, so Mainwaring sent a note to the divisional commander asking for orders, but the adjutant returned to say that he could find nobody in authority. As a result, Mainwaring gave orders to start withdrawing at 1730 and, moving slowly in 'artillery formation', the battalion reached Malincourt at dusk. By now his men were utterly exhausted, and he put them in a barn to rest.

A short time after dark Mainwaring, now in his third night without sleep, met the acting brigade-major of the 10th Brigade, who immediately attached a group of sixty men of the 2/R Warwicks and promised to send Mainwaring proper orders as soon as possible. At about 2100 Mainwaring, not having received the promised orders, went to find someone in authority but all he encountered was a cavalry regiment in the process of leaving the village, so he decided that he had no option but to continue the retreat and at about 2330 he led his men out of Malincourt, heading for Ronssoy. At 0300 the column arrived at a village which Mainwaring soon discovered was Beaurevoir, not Ronssoy. In the darkness he had taken a wrong turning. However, there was little he could do about it, so he gave the men two hours' rest in a barn (he got no sleep himself) and at 0500 they left, heading for Estrées, but with the intention of going on to St Quentin via Joncourt and Levergies. Then, at some time between 0500 and 0600, the 2/RDublin met the 1/RWarwicks on the road. Elkington, being the senior, now took command.

The 1/RWarwicks had had a generally similar experience but had become even more split up than the 2/RDublin. One company had been detached during the mid-afternoon of 26 August to escort some guns, while Major Poole, Elkington's second-in-command, was in charge of a second group which had also picked up two companies of Royal Irish Fusiliers as well as small groups from two other battalions. Poole's group never rejoined the main body of the battalion and later split up into a number of smaller parties, and although some men were captured, most escaped in a long march, reaching the River

Marne at Lagny on 5 September. The other two groups of 1/RWarwicks were Elkington's group and the sixty men who had latched on to the 2/RDublin.

The joint 1/RWarwicks and 2/RDublin group, numbering only between 250 and 300 men, now made its way slowly and painfully toward St Quentin. Behind them was the constant noise of German artillery and the route was littered with ditched baggage, stores and ammunition, all evidence of a hasty and ill-organized retreat.

As the column wearily approached St Quentin, Elkington sent Mainwaring on ahead. Once in the town one of the first people he met was the Corps Commander, Smith-Dorrien, who, having heard his story, advised him to get a train for the exhausted men. Mainwaring sent a message to Elkington with this news and suggested that they should meet at the station. Mainwaring then went to the station himself but found it deserted, and he was still there when Elkington arrived with the column at about 1400.

Mainwaring decided to make one more attempt and went to the Mairie with an interpreter. There he met the mayor, who was unable to arrange a train but who did agree to supply some food (which reached the men at about 1500). At this point a man entered with a note informing the mayor that the town was about to be attacked by Germans. The mayor then rounded on Mainwaring, telling him that the presence of British troops would cause the Germans to bombard the town, inevitably killing innocent women and children.

Returning to the column, Mainwaring discussed the situation with Elkington and they both then addressed their own battalions, explaining the situation and saying that they could either march on to Noyon, twenty-five miles away, or remain and in all probability be captured. Not one man in either battalion volunteered to march and both colonels concluded that their men were physically exhausted and that, for the time being at least, they could do no more. Elkington then directed the men to pile up their weapons, while Mainwaring returned to the mayor's office where he signed a paper stating that, since the men could march no further, he agreed to surrender them 'unconditionally'.

Mainwaring then went back to the station, but at about 1600 there came a new development. Major Bridges of the 4th Dragoon Guards had assembled a cavalry force from a number of groups and individuals whom he had found wandering around the battlefield and had organized them into an effective body, and on that afternoon they were screening St Quentin from the advancing Germans. On hearing that there were a number of infantry stragglers in the town Bridges dispatched a small party to round them up, but the captain in charge returned with the news that they were the remnants of two battalions with their officers and that these had already given a written agreement to surrender.

In the late afternoon Elkington and Mainwaring made renewed appeals to their men to move on, but when they again refused, Elkington found an abandoned horse and departed at about 1800. He collected stragglers as he went and eventually reached the 5th Division, where he got wagons for the exhausted men in this new group. He then rode on and reached the 4th Division at 1700 on 28 August.

Elkington's sudden departure left Mainwaring with the troops, but the situation changed yet again with the arrival of Major Bridges who had come into St Quentin to address the men and encourage them to start marching. Bridges was given such a hostile reception that he realized the men were at the end of their tether. Something desperate was needed to break them out of their despondency. In a flash of inspiration he obtained a penny-whistle and a tin drum from a nearby toyshop, and he and his trumpeter then marched around the fountain in front of the station playing the 'British Grenadiers' and 'Tipperary'. This utterly incongruous display so tickled the men that they were soon smiling and on their feet. The cavalry and infantry officers then started to get them into some semblance of order. Also during this period, Mainwaring and one of Bridges's officers went to the Mairie and retrieved the surrender document. Mainwaring led 400 men out of St

Quentin at 2100. They reached Noyon the following evening and from there they went by train to Compiègne.

Elkington and Mainwaring then reorganized what was left of their respective battalions and, after some sleep and food, both units were soon back to their original standard of efficiency and morale and took part in the Allied advance. After two days, however, the two colonels were suddenly arrested and arraigned before a general court martial on 12 September where they were charged with cowardice and conspiracy to surrender. At the trial both men were still very tired and were not allowed any help in their defence. Despite this, though both were cleared of cowardice, they were found guilty of the second charge. The court then gave the only sentence open to it. Both officers were to be cashiered – that is, stripped of their commission and deprived of any benefits, such as a pension.

Elkington and Mainwaring were sent back to England as civilians, where both got on with their lives, albeit in quite different ways. Mainwaring had not been in the best of health before going to France in 1914 and spent the remainder of his life in quiet retirement, although he became a popular and respected member of his local community. He died in 1930 at the age of 66.

Elkington, who was 48 at the time of his dismissal, adopted a quite different course by enlisting in the French Foreign Legion in October 1914. He served as a legionnaire under his own name and distinguished himself on numerous occasions before being seriously wounded on 28 September 1915. He was in hospital for many months and then, having lost the use of his right leg, was discharged and returned to England. He had been awarded the Croix de Guerre and the Médaille Militaire while in the Legion. When his story became known he was reinstated in the British army in his previous rank and seniority by King George V and shortly afterwards was awarded the DSO. He was, however, no longer fit for military service and retired, living until 1944. As far as is known, Mainwaring never made a formal attempt to achieve reinstatement, not even when Elkington was exonerated and awarded the DSO.

In August 1914 military operations were little changed from those of the Napoleonic Wars, apart from the use of railways for strategic moves and the introduction of the machine-gun at the tactical level. Thus, the principal arms were foot infantry and mounted cavalry, supported by field artillery. The last used breech-loading guns, an obvious advance over muzzle-loaders, but they were still horse-drawn and used direct fire. Communications were the responsibility of the Royal Engineers, but the means were rudimentary, and while there were limited stocks of telephones and telegraph instruments, there was no question of cable being laid during a mobile battle. Thus – as at Waterloo a century before – communications were confined to mounted officers, such as ADCs, and runners. In addition, intelligence staffs were virtually non-existent, and in the hectic days of August 1914 battalion commanders frequently had little or no idea of the location of either their subordinate companies or their superior HQs, and knew nothing of what the enemy was up to, apart from what they saw of those with whom they were in direct contact.

At the time, the higher echelons of command were in disarray. Smith-Dorrien, the Corps Commander, appears to have been able to exercise some sort of control up until the evening of the Battle of Le Cateau, but thereafter, as the withdrawal progressed, the chaos and confusion increased and – at least at battalion level – nobody seems to have known what was going on. There was no information and there were no orders.

The 10th Brigade headquarters became ineffective – at least so far as Elkington and Mainwaring were concerned – from about midday on 26 August, and repeated attempts to find it failed, although they occasionally bumped into individual staff officers. Brigadier-General Haldane's activities throughout these events are not at all clear and he certainly did not – as his duty surely required him to – visit either the 1/RWarwicks or 2/RDublin at a time when both commanding officers were desperately in need of orders and guidance.

In particular the senior staff failed to make any effort to

obtain a train for the two battalions. Smith-Dorrien was in St Quentin when Mainwaring arrived, and some staff officers had certainly been in the town that morning arranging transport for other units.[3] It should, therefore, have been the business of a staff officer from II Corps or the 4th Division to remain to the end, finding trains or other means of transport for the exhausted troops. It should certainly not have been left to the tired and confused Mainwaring.

There is no question that Elkington and Mainwaring were men of great experience who were dedicated to their profession and who had the reputations of their battalions and the interests of their men uppermost in their minds. But nothing in the experience of either man had prepared them for the events of that first week in France, where the rapidity of events, frequent changes in orders – or the total absence of them – and the dismal aura which always attaches to a retreat had a detrimental effect on these two middle-aged men. By the time they reached St Quentin, both commanding officers were clearly physically and mentally exhausted, as were all their officers and men, but in these two particular cases, they bore the added responsibility of command.

They had been in the same brigade since February 1914 and would undoubtedly have met on exercises and at conferences. Following the declaration of war, both men travelled to Yorkshire together, shared the same barracks for a short period, and then travelled south again and crossed the Channel on the same ship. Thus, while there is no evidence that they were close friends, they undoubtedly knew each other reasonably well.

Mainwaring should not have signed the document in the Mairie, but there can be no doubt that he intended nothing dishonourable by it, the factors uppermost in his mind being the exhausted state of his troops and the danger to French civilians in St Quentin, especially the women and children.

One curious event was Elkington's departure from St Quentin at 1800 on 27 August. He later explained that he had decided that the 'danger was past' and that he had been 'anxious to get on and pick up stragglers'. There is no question that Elkington was a courageous man with a strong sense of duty, but quite what motivated him to leave the remnants of his battalion with Mainwaring is difficult to explain, except, perhaps, that as a result of exhaustion and the pressure of events, he was acting neither rationally nor in character.

Apart from the two commanding officers, there were no other courts martial for the events involving the two battalions at St Quentin. Furthermore, it appears that none of the other officers or soldiers of the two battalions ever blamed their two commanding officers, nor did either Elkington or Mainwaring ever utter one word of blame against their men for what had happened – not even their refusal to march any further during the afternoon of 27 August. It is also worth noting that Brigadier-General Haldane sacked the commanding officer of the Royal Irish Fusiliers during the retreat from Mons, thus losing three out of four of his infantry battalion commanders within the space of a week.

Finally, it is significant that Field Marshal Viscount Montgomery, who was a lieutenant with the 1/RWarwicks in August 1914 (but who was not at St Quentin as he was with Major Poole's group), remained a steadfast friend of Elkington and his family for the rest of his life. Montgomery was noted for his stern sense of duty and morality, and it is unlikely that he would have remained so close to Elkington had he felt that the latter had failed in any way.[4]

COMMANDER X

The Gulf War, 18 January 1991

On 17 January 1991, the day Gulf War hostilities started, USS *Nicholas* (FFG-47), a Perry-class frigate, arrived in the northern Arabian Gulf in company with two Kuwaiti fast patrol boats, *Istiqlal* and *Al Sanbouk*. *Nicholas*'s commanding officer was Commander X –, US Navy, who was also in tactical control of the two Kuwaiti units, and he was under the tactical control of Commander Destroyer Squadron Thirty-five (Comdesron 35).[5] X—'s primary mission was to rescue downed Allied aircrew, it being anticipated that aircraft losses would be high and that his group could be undertaking up to fifteen combat search and rescue missions per day. To achieve that, *Nicholas* had embarked four helicopters – two US Navy Sikorsky SH-60s and two US Army Bell OH-58Ds – but aside from them she had no air cover.

The major perceived that the threat to the *Nicholas* group was from Iraqi-launched anti-ship missiles, for the ships' deployment area was well within range of two known batteries of land-based, Chinese-supplied Silkworm missiles, while Exocet missiles launched from aircraft could approach from almost any direction.

The effectiveness of the French-made air-launched Exocet was fully appreciated in naval circles. The sinking of the British destroyer HMS *Sheffield* and the transport *Atlantic Conveyor* in the Falklands War in 1982 by Exocets had been widely studied by all navies at the time. The same missile was also used by the Iraqi airforce and had been responsible for causing severe damage to the frigate USS *Stark* (a sister-ship to *Nicholas*) and numerous tankers in the Gulf in the late 1980s. The other missile, the Chinese-made Silkworm, was a copy of the Soviet SS-N-2 Styx, which had sunk the Israeli destroyer *Eilat* in 1967, while the Silkworm itself had been used in the confrontations around the Paracel Islands in the South China Sea in the 1980s.

TABLE 12: *Iraqi Anti-ship Missiles*

	Silkworm HY-2	*Exocet AM-39*
Country of origin	China	France
Power	Rocket (liquid fuel)	Rocket (solid fuel)
Launch	Land	Aircraft
Speed	Mach 0.9	Mach 0.93
Maximum Range	60 miles	44 miles
Warhead	1,131lb	364lb
Homing	Active radar	Active radar

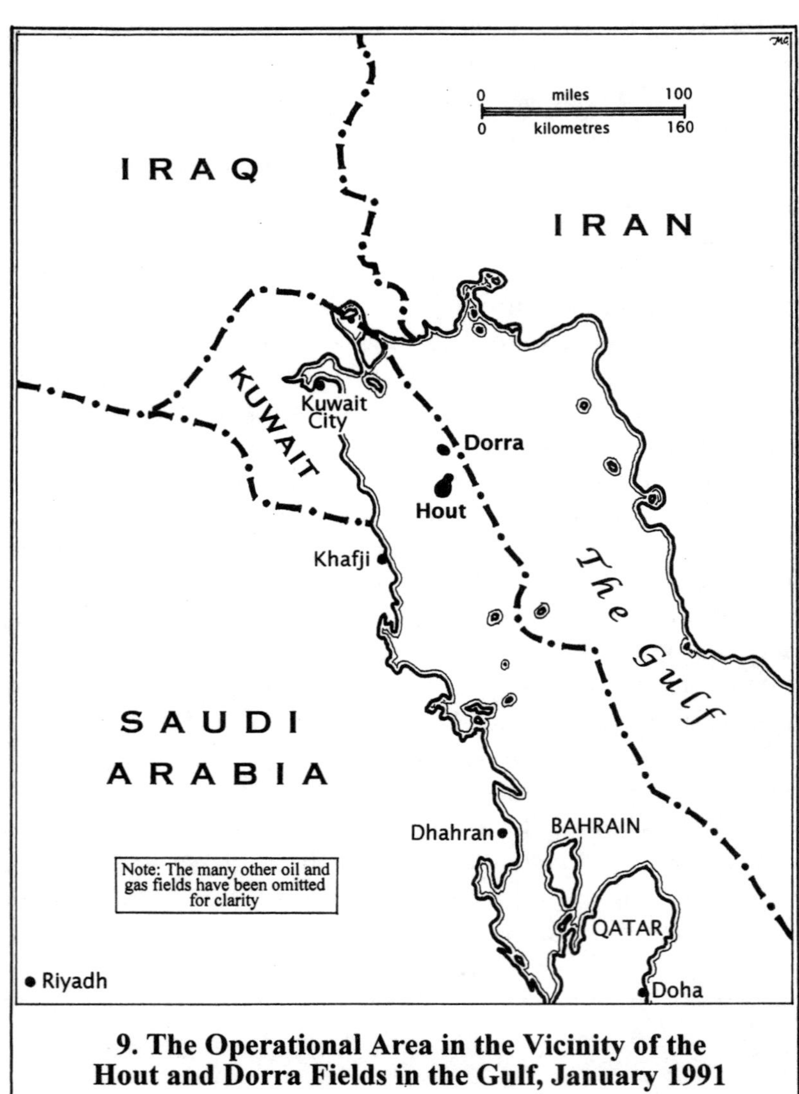

9. The Operational Area in the Vicinity of the Hout and Dorra Fields in the Gulf, January 1991

Thus, both types of missile posed a credible, serious and well-understood threat to Commander X—'s three ships.

The most significant physical features in X—'s allotted area were two groups of Kuwaiti-owned platforms, one in the Hout oil field, the second in the Dorra gas field. It was known that they were unmanned in peacetime, but it was anticipated that they might be used by the Iraqis as forward observation posts to acquire maritime targets for the missiles. It was therefore essential that the platforms be investigated.

Through no fault of X—'s, his group was twenty-four hours late in deploying, a delay which, combined with the outbreak of hostilities, greatly increased the sense of urgency. The three ships approached the oil platforms before last light on 17 January 1991, and as soon as they were within range X— launched the two Army OH-58D helicopters to reconnoitre the platforms. They went first to the Hout field, where nothing suspicious was seen, and then moved on to the Dorra field, where the aircrew saw a few men and 'possible' antennae on some of the platforms. At about this time lookouts aboard the three ships reported seeing tracer fire coming from the area of the platforms and the reports were passed to *Nicholas*'s command centre, but, owing to a misunderstanding, the reports were thought to have come from the helicopters.

X— considered it essential for the safety of his group that he eliminate the threat posed by the men on the platforms, so he withdrew his ships a few miles to the south while he sought permission for such an operation from Comdesron 35. The request was approved, but with the proviso that the plan for the operation should be passed to Comdesron 35 before the attack started.

A further reconnaissance of the Dorra platforms was then carried out by one of *Nicholas*'s Navy SH-60s in order to provide firm information for the attack, during which the helicopter flew close to each platform in turn. The helicopter crew was able to confirm that of the nine platforms, seven appeared to be occupied and that on the top decks of six of them were canvas-covered devices, one of which was identified as a 'definite' Soviet-style 23mm AA gun, the other five as 'possibles'. This was a serious addition to the threat, since the 23mm guns had a maximum range of 7,700 yards and a firing rate of 1,000 rounds per minute, and were thus a danger to the ships and, in particular, to the helicopters. The helicopter crews also saw military equipment, including Zodiac inflatable dinghies, outboard motors, sandbagged emplacements, diving equipment and cases of ammunition. People dressed in green uniforms were seen on five platforms. Some of them simply watched the helicopter, but a few waved, while on two separate rigs a single man was seen to wave what appeared to be a white cloth.

On its return to *Nicholas* the helicopter's report was logged and the captain informed, but he made it clear that Comdesron 35 was not to be told about the 'white flags' and that the attack would continue as planned. As happens aboard a warship, news of the 'white flags' was quickly passed around and two sailors spoke to the executive officer, saying that, since the Iraqis appeared to be prepared to surrender, they hoped the proposed attack would be cancelled, a view which the executive officer passed to the captain. The captain replied firmly that he had sole responsibility for the safety of his ship and that the waving of white cloths by two men on two separate rigs did not constitute a surrender by all personnel on all platforms, nor did it even indicate that all the personnel on either of those two platforms intended to surrender. Commander X— did, however, consult the Kuwaiti commanding officers. Both told him that during the Iran-Iraq War the Iraqis had on several occasions used white flags as a ruse and then, once their opponent had been lulled into a false sense of security, they had opened fire.

A second naval officer who had heard about the 'white flag' incident also spoke to the captain about it, and he and the executive officer both observed that X— 'became agitated' and repeated that this information was not to be passed up the chain of command. X— considered his mission to be time-critical and expected to have to conduct large numbers of search and rescue missions. Though he considered the possibility of firing warning shots at the platforms, he rejected such a scheme since it appeared that the Iraqis were in radio contact with the shore and thus able to target anti-ship missiles against his force. He therefore instructed the executive officer to proceed with the attack plan.

The following morning (18 January) a briefing was held aboard *Nicholas*, following which the operation was launched. At the outset the two Army helicopters attacked the two southern platforms of the Dorra field with rockets and machine-guns. Following this, *Nicholas* attacked the four central platforms with fire from her 76mm gun at a range of 4 to 5 miles, while *Istiqlal* attacked the remaining three northern platforms with her 76mm and 40mm guns, as well as her 0.50-inch machine-guns, suggesting that the range was much shorter.

Resistance by the Iraqis was, at best, limited. Some machine-gun fire was reported and a rocket grenade was launched at *Istiqlal*, but it fell far short of its target. As soon as it was clear that resistance had ceased the bombardment was terminated and the ships closed on the platforms. They took off nineteen unwounded men, four casualties, who were given immediate medical treatment, and the bodies of five killed. Among the equipment found on the platforms were four 23mm guns, each with about 1,000 rounds of ammunition, AK-47 rifles, SA-5 shoulder-launched anti-aircraft missiles, UHF/VHF radios and two HF radios fitted with encryption devices.

There matters might have rested, but two weeks later a member of *Nicholas*'s crew wrote to a superior officer claiming that the attack had been carried out despite the sighting of white flags, and this led to an official US Navy inquiry being convened. This sat in June 1991 and, after hearing evidence from all concerned, concluded that Commander X— had been legally justified in carrying out the attack but that he had shown a lack of judgement in his failure to notify higher command of the sighting of the white flags and in his failure to request advice.

X— had been instructed to report his attack plan to superior authority. He failed to do so. Nor did he mention the 'white flag' incidents but instead instructed his subordinates that they were not to do so either. He was subsequently severely criticized for this on the grounds that had he done so, he could have been given advice on what was a difficult legal situation. That he did not do so, however, was entirely an internal US Navy administrative matter and did not affect the legality of his decision to fire on the platforms.

The combination of a known Iraqi anti-ship missile capability and the presence of Iraqi troops aboard the oil platforms posed a serious and indisputable threat to Commander X—'s force. Further, it was logical for him to deduce that the latter's mission was to observe Allied naval activity and to report back to shore by radio, probably with the intention of providing target data for a subsequent attack by anti-ship missiles. Indeed, there was no other credible reason for the Iraqi troops to be on the platforms.

Commander X— had a primary responsibility for his ship and the safety of its crew, as well as for the crews of the two ships under his tactical direction. In addition, he had been tasked with search and rescue missions which, at the time of the oil rigs incident, he had every reason to believe might amount to perhaps fifteen rescues per day. There can, therefore, be no doubt that he was entitled to eliminate such a direct threat to the completion of his mission and to the safety of his ship and crew.

Had the threat come simply from the weapons on the platforms, then X— could have ordered his three ships and four helicopters to remain outside the effective range of the most capable weapons, which in this case were the 23mm guns. This would have been a nuisance but would not have compromised his search and rescue missions, unless airmen fell within the range of the guns. The real danger, however, came from the Silkworm and Exocet missiles. Since radio antennae had been positively identified on the platforms, it was a reasonable deduction that the Iraqis on the platform were in contact with the shore and that they were capable of acting as target spotters for either shore- or air-launched missiles. Indeed, the platforms had no other tactical significance.

One or two men waving white flags may indicate that they, as individuals, have no further wish to fight, but that does not mean that the remainder do not wish to do so either. In a land battle it may well be that an individual wishing to surrender can run towards an enemy position waving a white flag and have a reasonable expectation of being allowed to give himself up, but such an action would not necessarily indicate that the entire unit wished to do so. At sea, however reluctant certain individuals may be to fight, there can be no question of an opponent accepting their surrender while the commander and the majority of the crew wish to fight on. In this particular incident, army troops aboard a platform at sea did not fit neatly into either traditional pattern, but it seems reasonable to suppose that the minimum requirement for demonstrating a positive wish to surrender would have been a large white flag, either hoisted above the top deck of the platform or laid out on the top deck. Nor should it have been difficult for the Iraqis to have established radio communication with either the US or the Kuwaiti ships had they wished to surrender.

There must have been an officer in charge of the Iraqi positions, but X— had no means of knowing which of the platforms he occupied, and the helicopter reconnaissances saw nothing to suggest the particular location of a headquarters. Certainly, no effort was made by any individual on the platforms to identify himself as the commander and offer to surrender.

The commanding officer of USS *Nicholas* indisputably acted correctly in the circumstances, but his dilemma indicates the type of complex problem which can face a commanding officer in modern operations.[6]

9

Suicide

The concept of suicide as an explanation for personal failure by a commander has been deeply rooted in Japanese culture for many centuries and has taken a variety of forms. One method was deliberately to seek death through a solitary plunge into the enemy ranks, in which a small number of enemy might be killed but personal death was reasonably assured. This course was adopted by Yamamoto Kansuke, the battlefield commander for Takeda Shingen, who in 1561 devised a plan to surprise the enemy at the fourth battle of Kawanakajima. When it became apparent that his plan had failed, he rushed into the enemy ranks seeking death, although in the event he fought so ferociously that, despite many wounds, he remained alive and then had to kill himself with his sword. During a siege in which the garrison was eventually forced to negotiate a surrender, one of the victor's usual conditions was that the castle commander should commited *seppuku* (ritual suicide). One of many examples is that of Kikkawa Tsuneie, who commanded the garrison at Tottori castle when, in 1581, it was besieged by the great warrior Toyotomi Hideyoshi. The garrison held out for almost seven months and was reduced to the extremes of privation in the process, but Hideyoshi demanded Kikkawa's life as a condition of the surrender.

The great majority of Western commanding officers feel every bit as strongly about their responsibilities as did the

Japanese samurai, but only a very few have taken matters to such an extreme that they no longer feel able to go on living.

LIEUTENANT-COLONEL BEVEN

During the British Army's five-year campaign in the Iberian Peninsular (1809–14) the Duke of Wellington exercised close supervision over his troops. It is therefore not surprising that when, from time to time, his commanding officers made mistakes he should have been aware of them. One such occasion arose on 10 May 1811 when, following Wellington's defeat of Marshal Masséna at Fuente de Oñoro, the British 5th Division surrounded a French garrison of 1,300 men in the town of Almeida. Wellington warned the divisional commander and his brigade and commanding officers to 'keep a sharp lookout', but despite this the French managed not only to escape in the darkness but also to blow up the town as they departed. The British, somewhat belatedly, gave chase and killed about 150 of the French rearguard, but Wellington, irate that his warnings had been disregarded, issued a public rebuke to the regiments which, in his opinion, were responsible for the lapse in attention. One of the units concerned was the 4th Regiment of Foot whose commanding officer, Lieutenant-Colonel Beven, had just returned from England where he had recently got married. Beven took strong exception to Wellington's criticism of his regiment and wrote to the Duke setting out his version of events. Wellington rejected the letter, remarking that 'it had been a sad, bungled business, wherever the fault might lie, and that Colonel Beven's letter was no excuse at all'. On hearing this, Beven retired to his hut, wrote a letter of farewell to his regiment and shot himself.

Lieutenant-Colonel Beven clearly thought that his regiment had been deeply and publicly humiliated by the Duke of Wellington's comments and felt that he must shoulder the blame in an equally public way; and it is to be presumed that not even the thought of his recent marriage deterred him from his course. The whole affair was observed by a 21-year-old staff officer, Captain Thomas Browne, who, having described the events, went on to comment that

> it is hard indeed if a General commanding cannot, at his discretion, use terms of censure or of praise on actions which may have caused him disappointment or gratification in his military arrangements. Officers of superior rank were included in the same disapprobation as Lieut. Col. Beven, and yet they bore it calmly, and in a very few weeks nothing more was thought of the affair.

Browne's views would seem to have been more perceptive and balanced that those of Beven.[1]

KAPITÄNLEUTNANT PETER ZSCHECH

There were two known suicides among U-boat commanders during the Second World War, one of them being Kapitänleutnant Peter Zschech, who took command of *U505*, a Type IXC, in September 1942 at the age of 24. His first patrol in command started on 4 October 1942 and he scored his first success, sinking a merchant ship off Trinidad, on 7 November. On 10 November, however, he was surprised on the surface by an RAF aircraft which caused considerable damage and wounded several men, and it was only after strenuous efforts by the crew that a limited diving capability was restored. Zschech had to rendezvous with three other U-boats during the return voyage, first to obtain more morphine for the wounded crewmen, then to obtain spares and, finally, to refuel and obtain medical advice. Eventually, with all aboard exhausted, *U505* reached Lorient on 12 December.

Following lengthy repairs, *U505* sailed again on 1 July 1943, but several faults were discovered and Zschech returned for them to be solved. He put to sea again on 3 July but was attacked twice on 8 July, first by aircraft and then by ships, which forced him to return yet again, reaching Lorient on 13 July. He then sailed on three separate occasions (1, 14 and 21 August), but each time he was forced to return to port within twenty-four hours because of strange noises when the boat descended below a depth of about 50 metres.[2] Zschech sailed again on 18 September. On 23 September his boat was attacked by an Allied aircraft and he crash-dived, in the course of which the main ballast pump, which was essential to the proper functioning of the boat, was so severely overstressed that yet another return to base was necessary. *U505* sailed again on 9 October and this time got well clear of the Bay of Biscay, but on 23 October she was subjected to a very accurate depth-charge attack by surface ships. She survived the attack, but the crew then discovered that their captain had retired to his cabin and shot himself in the head, dying instantly and leaving his second-in-command to save the boat and her crew.[3]

KAPITÄNLEUTNANT HORST HÖLTRING

The suicide of Kapitänleutnant Horst Höltring was an even more melodramatic affair. Höltring, aged 30, was in command of *U604* in July 1943 when she was attacked on three separate occasions, incurring such serious damage that he was forced to rendezvous with two other U-boats (*U172* and *U185*), and divide his crew between the two. *U604* was then scuttled. Höltring was with part of his crew aboard *U185* when, on 24 August, she was attacked by two US Navy aircraft and so badly damaged that she began to sink. When water reached the battery, chlorine gas spread through the submarine, where-upon the order to abandon ship was issued. Höltring rushed to assist two of his crew in the fore-ends who had been so badly wounded in *U604* that they were unable to move. When it proved impossible to carry them out, they pleaded with Höltring to put them out of their misery. He did so, shooting each in turn with his service pistol, and then he shot himself, dying instantly.

Kapitänleutnant Wolfgang Wenzel

One attempted suicide was somewhat bizarre. This occurred when *U231*, a Type VIIC, was sunk by depth-charges from a British aircraft on 13 January 1944. Seven men were killed and the commanding officer, Wenzel, gave the order to abandon ship. He, however, remained standing on the bridge and, when the last man had left, he took his pistol, placed the muzzle in his mouth and pulled the trigger. The weapon went off, but the bullet lodged in the back of his neck without killing him, and his crew returned to save him.

Zschech, Höltring and Wenzel appear to have been over-whelmed by events that would have daunted almost anyone, and yet they had continued to go to sea. Indeed, it is curious that in Zschech's case Admiral Dönitz, who normally kept a very careful eye on his commanding officers, should have felt it necessary to keep him in post. Dönitz was by no means as inflexible a commander as some commentators have suggested and on several occasions he relieved officers who needed a rest without fuss, either sending them to a training establishment or, in a few cases, returning them to the surface navy.

One commanding officer, Heinrich Bleichrodt, had a very exacting career in command of *U109* but on 27 December 1942, when in the Caribbean, he decided he had had enough and signalled Dönitz that he was ill and was returning to base. Dönitz sent several signals ordering him to continue with the patrol and, if necessary, to hand over command to his executive officer; but to no avail, and *U109* continued to head for home. Three days were then wasted in an unsuccessful attempt to transfer *U109*'s torpedoes to other boats which had run low, but eventually they gave up and *U109* docked on 23 January

1943. Dönitz might well have treated Bleichrodt very harshly for a totally wasted patrol. Instead he sent him to a U-boat training job and promoted him to Korvettenkapitän before the war's end. It is possible that Zschech might have been treated in the same way and that the worst fate he would have suffered would have been a posting to the surface navy.[4]

Captain John Philip Cromwell

In 1943 the Japanese sought to reduce their increasingly heavy surface ship losses to submarines by forming convoys, and the US Navy sought to counter this by forming 'Co-ordinated Submarine Attack Groups' with a senior officer aboard one boat as the overall commander. The first such group sailed in November 1943 under the command of Captain Cromwell, an experienced US Navy submariner, aboard USS *Sculpin*. On 19 November *Sculpin* was attacked by a Japanese destroyer and forced to the surface where she came under such heavy gunfire that the submarine's captain, Commander F. Connaway, ordered the ship to be abandoned. Cromwell, however, was privy to the highly classified Anglo-American Ultra system and had recently also been involved in planning some top-secret operations which had not yet taken place. Faced with the starkest of choices, he decided that he could not guarantee to keep safe the secrets with which he had been entrusted if subjected to a ruthless interrogation. He therefore calmly informed Connaway that he would remain with the boat. Quite why he should have been allowed to place himself in such jeopardy in the first place is not clear, but he nevertheless reacted in a most courageous way.[5]

Kapitän zur See Hans Langsdorff

Hans Langsdorff was the commanding officer of the pocket battleship *Admiral Graf Spee*, in which he carried out a commerce raiding campaign in the South Atlantic in August–December 1939. After a number of successes, Langsdorff decided to carry out one last raid against the heavy traffic off the mouth of the River Plate before returning to Germany. His men deserved some leave and his ship desperately needed to undergo repairs. Unfortunately for him the British and French navies had deployed an enormous force to look for him, which included a small group operating off the mouth of the River Plate.

The lone German ship encountered three British warships early on 13 December 1939. Two of them had single funnels and so were mistakenly identified as destroyers, and Langsdorff headed straight for them. By the time the mistake was realized Langsdorff was already committed to the engagement, in which he inflicted heavy damage on one British cruiser but also suffered damage to his own ship. This was so severe that he decided that he had no option but to make for the Uruguayan port of Montevideo to carry out essential repairs before commencing the long and dangerous voyage home.

However, isolated from friendly naval support, and lacking the means to replenish supplies (particularly of ammunition) and to effect repairs, he found that he had placed himself in a strategic and diplomatic trap. Faced by ever-increasing British naval forces off the mouth of the Plate, he took his ship, now manned by a minimal crew, to a point just outside territorial waters where delayed-action fuzes were set. Then the last elements of the crew transferred to waiting boats.

The entire crew of the *Graf Spee* were taken to Buenos Aires where they were interned. Langsdorff also went to Buenos Aires, where he ensured that his crew were as well looked after as circumstances permitted, but when the Argentine government insisted that the crew be interned rather than treated as distressed seamen, he realized that he could do nothing more to help them and shot himself. Before doing so he left a letter to the German ambassador (but in effect intended for Hitler) which is so clear in its statement of responsibility and so dignified in its tone that it merits setting out in full:

Your Excellency, After a long struggle I reached the grave decision to scuttle the pocket battleship *Admiral Graf Spee*, in order to prevent her from falling into enemy hands. I am still convinced that under the circumstances, this decision was the only one left, once I had taken my ship into the trap of Montevideo. For, with the ammunition remaining, any attempt to fight my way back to open and deep water was bound to fail. And yet only in deep water could I have scuttled the ship, after having used the remaining ammunition, thus avoiding her falling to the enemy.

Sooner than exposing my ship to the danger that after a brave fight she would fall partly or completely into enemy hands, I have decided not to fight but to destroy the equipment and then scuttle the ship. It was clear to me that this decision might be consciously or unwittingly misconstrued by persons ignorant of my motives, as being attributable entirely or partly to personal considerations. Therefore I decided from the beginning to bear the consequences involved in this decision. For a captain with a sense of honour, it goes without saying that his personal fate cannot be separated from that of his ship.

I postponed my intention as long as I still bore responsibility for decisions concerning the welfare of the crew under my command. After today's decision of the Argentine Government I can do no more for my ship's company. Neither will I be able to take an active part in the present struggle of my country. I can now only prove by my death that the men of the fighting services of the Third Reich are ready to die for the honour of the flag.

I alone bear the responsibility for scuttling the pocket battleship *Admiral Graf Spee*. I am happy to pay with my life for any possible reflection on the honour of the flag. I shall face my fate with firm faith in the cause and the future of the nation and of my Führer.

I am writing this letter to Your Excellency in the quiet of the evening, after calm deliberation, in order that you may be able to inform my superior officers, and to counter public rumours if this should become necessary.

(Signed) Langsdorff
Captain
Commanding Officer of the sunk
Armoured ship *Graf Spee*[6]

Langsdorff was a man with a sense of honour who appreciated that the destruction of the *Graf Spee* would adversely affect the reputation of the German navy. One of his finest obituaries came from the British naval attaché, who covered both Buenos Aires and Montevideo, and who had played no small part in frustrating Langsdorff's efforts to obtain assistance from the Uruguayan authorities. This came in his report to London:

Since starting to draft this letter the news has been received of the suicide of Captain Langsdorff. I attribute his preferring to shoot himself after the event rather than being blown up with his ship to his principle that it was his duty to be with his ship's company until after every possible provision had been made for them in Argentina. He was obviously a man of very high character and was proud of the fact that he had not been the cause of a single death as the result of any of his captures.[7]

Langsdorff remains one of the very few commanding officers to have committed suicide who have enhanced their reputation by having done so.[8]

Suicide is the action of a person who has, for a variety of reasons, reached the end of his tether, and it has been comparatively rare among Western naval and military commanding officers. Those who have committed suicide seem to have done so for one of four reasons. Zschech and Höltring did so when the stress of combat became too great, probably the most common reason. Beven is an example of those who believe that they are either directly or indirectly responsible for their unit failing in its duty and being publicly reprimanded. Wenzel seems to have been a unique case, although he may have been endeavouring to join those captains who have gone down with their ship, which is discussed in more detail in the following chapter. Cromwell seems to have been both a unique and an unfortunate case in that he should not have been at sea in the first place and his selfless action actually suggests that he would probably have had more than enough courage to have resisted the Japanese anyway. Finally, there is Langsdorff who was trying to expiate his own mistaken decision to enter Montevideo, which had been taken with the very best of motives.

10

The Ultimate Demonstration of Responsibility 'The Captain Went Down with His Ship'

One of the most extreme demonstrations of a commanding officer's responsibility for the unit under his command is the naval practice – if that is the right word for it – of a captain 'going down with his ship'. No proper definition of this particular action can be found, but it can be taken to mean that, his ship having suffered some disaster either in battle, by the action of the sea, by fire, by neglect or by the action of another ship (for example, in a collision), the captain, who is physically capable of attempting to save himself, voluntarily chooses to remain with his sinking ship and so meets his death.

Such acts have been recognized and accepted among a number of seafaring nations for some time, but a detailed search has failed to identify more than one example before about 1850 and, despite considerable research, it has proved impossible to discover how and from where the idea originated. Indeed, although there are many descriptions of captains actually taking this course, there seems to be no literature which attempts either to give a moral justification for it, or to criticize it. Furthermore, there are no official regulations containing even a hint that this is a desirable course of action. It seems, therefore, to be a matter of international naval folklore that has been passed on by word of mouth.

In the era of wooden warships, which endured well into the latter half of the nineteenth century, it was extremely rare for a ship actually to sink in battle. Most engagements ended with a boarding action, culminating in the defeated captain surrendering his sword on his quarterdeck, following which he was treated as an honoured prisoner-of-war. Occasionally, the ship would be set on fire or a chance shot would reach the ship's magazine, whereupon the ship would blow up, as happened to the French *L'Orient* at the Battle of the Nile in August 1798, but in such circumstances the captain had no hand in his fate. The only exception that has been found to this general rule concerning the wooden ship era is that of Captain Archibald Douglas, who commanded HMS *Royal Oak* at the time of the Dutch naval attack on the British fleet in the River Medway in June 1667. His ship was set alight by enemy action. When it was clear that there was no possibility of saving it, Douglas ordered his crew to seek safety ashore, but, since the last order he had received had been to 'maintain his post to the last extremity', he himself remained aboard, telling the crew that 'it should never be said that a Douglas had quitted his posted without orders'.[1]

When ships were lost at sea, it was usually as a result of storm, fire or running aground, and in such circumstances captains sometimes lost their lives. However, this was almost always as a direct result of the disaster rather than a deliberate act by the captain, and no example has been found of a captain going down with his ship to demonstrate his responsibility for what had happened. Indeed, captains accepted that their duty was to their crew. To take one example among many, when Captain Edward Edwards, a man with a particularly rigid view of the duties of a captain, ran HMS *Pandora* ashore in Endeavour Straits in the Dutch East Indies in 1791, he survived and led his crew back to civilization; the question of any other course of conduct never arose.

The concept of the 'captain going down with his ship' seems to have crept in at some point in the middle of the nineteenth century and, at least originally, it seems to have applied to both naval and merchant services. One of the earliest examples is the captain of the liner RMS *Amazon*, who went down with his blazing ship on 4 June 1852, while the most famous is Captain

Edward Smith of RMS *Titanic* in 1912. Then too, some of the survivors of the troopship *Khedive Ismail* which, as described in Chapter 4, sank in 1944, told how they saw the Master, Captain Whiteman, calmly standing on the bridge as the ship capsized.

One of the earliest examples involving a naval commanding officer is that of the USS *Oneida*, a wooden-hulled steam corvette. While on passage off the Japanese port of Yokohama on 24 January 1870 she collided with the P&O liner SS *Bombay*. The American ship immediately started to sink and the commanding officer, Commander Williams, ensured that the crew reached the boats but then remained aboard, stating that it was his duty to 'go down with his ship'.

On 16 December 1900 a German naval training ship, *Gneisenau*, with 460 people on board, was lying at anchor off the port of Malaga in Spain when a gale developed. The ship was blown on to the outer breakwater and sank. The commanding officer, Kapitän zur See Kretschmann, was seen in the water by a Spanish sailor standing on the breakwater who threw him a rope, but Kretschmann refused to take it, stating that it was his duty to go down with his ship. After throwing his sword to his would-be rescuer, he did so.

There are numerous examples from opposite sides in the Second World War. On 8 June 1940, two British destroyers, HMS *Acasta* and HMS *Ardent*, were escorting the aircraft-carrier HMS *Glorious*, which was returning to Scotland after the unsuccessful Norwegian campaign, when they encountered the German battlecruisers *Gneisenau* and *Scharnhorst*. After a short, sharp engagement, first *Ardent* was sunk and then *Glorious*, both with heavy loss of life. Commander Glasford in *Acasta* then attacked out of a smokescreen and scored a close-range torpedo hit on *Scharnhorst*, but his ship came under devastating fire and was soon sinking. Of the crew of 138 just one man survived, who later described the last moments of his ship:

> when I was in the water I saw the Captain leaning over the bridge, take a cigarette from a case and light it. We shouted

to him to come on our raft, he waved 'Goodbye and good luck' – the end of a gallant man.

The British-owned Yangtze transport *Li Wo* was requisitioned for naval service in December 1941 as HMS *Li Wo* and given an armament of a single 4-inch gun and two machine-guns, while her civilian master, Tam Wilkinson, was commissioned as a lieutenant in the Royal Navy Reserve (RNR) and appointed commanding officer. Wilkinson sailed from Singapore on 13 February 1941 and the following day sighted a convoy of Japanese amphibious ships, escorted by a cruiser and several destroyers, en route to invade Sumatra in the then Dutch East Indies. Despite being very heavily outnumbered, Wilkinson headed straight for the enemy, engaging and damaging the nearest ship, until, with no more ammunition remaining, he rammed it. Having freed his ship, Wilkinson then came under fire from the Japanese warships but was able to survive for an hour before *Li Wo* began to sink. Realizing that no more could be done, he gave the order to 'abandon ship'. He was last seen by his crew standing on the bridge as his ship, its White Ensign still flying, disappeared below the waves. Wilkinson had never received any naval training but clearly had a firm understanding of what he perceived to be the traditions of the service that he had joined.[2]

At least a dozen Second World War U-boat commanding officers went down with their boats, and of these Korvettenkapitän Georg von Wiliamowitz-Möllendorf was typical. Born in 1893, he served in the Imperial German Navy throughout the First World War, joining the U-boat arm in 1917. He returned to the service in 1940, in which his second command was *U459*, a Type XIV tanker, and he proved to be one of the most successful and long-lived commanders of these ill-fated boats. *U459* shot down one RAF aircraft on 30 May 1943 and a second on 24 July, but on this occasion the aircraft crashed on to the U-boat's upper deck and then slid into the sea, leaving an unexploded depth-charge aboard the U-boat as it did so. Not realizing that it was still active, the crew pushed the depth-charge into the sea, where its fuse was activated as it passed under the U-boat's stern. The resulting explosion made it impossible for the boat to dive. Another RAF aircraft then appeared and attacked, causing so much damage that the captain ordered his crew to abandon ship. Having seen forty-one of his men safely into their dinghies, von Wiliamovitz-Möllendorf waved to them, saluted and then went below to open the sea-cocks and scuttle his boat. He was never seen again.

Captains of surface ships also went down. On 27 May 1941 two sailors aboard the doomed German battleship *Bismarck* sensed that Kapitän zur See Lindemann intended to go down with his ship and tried to force him into the sea, but he ordered them off and was clearly seen on deck saluting as his ship went down. Similarly, Kapitän zur See Hintze of the *Scharnhorst* was seen, unwounded, on the bridge as his ship went down on 26 December 1943.

Naval regulations give no evidence of any encouragement of this practice. The British Queen's (or King's) Regulations and Admiralty Instructions were reissued from time to time, but the regulation covering the commanding officer's responsibilities on the loss of his ship was quite clear, that of 1862 being typical:

> If one of Her Majesty's ships be wrecked or otherwise lost or destroyed, the Captain is to use every exertion to preserve the lives of the Crew; and when they, or as many of them as possible, are saved, he is to use his utmost endeavours to save the stores, provisions, and furniture of the Ship . . . He is to keep the Crew together, and to be very particular in preserving regular and perfect discipline among them.[3]

The whole emphasis of this instruction is on the survival of the commanding officer, his continuing responsibility towards his crew, and his duty to preserve documents, secret orders, accounts and so on. While the involuntary death of the commanding officer clearly cannot be ruled out, nowhere is there even the remotest suggestion that he should go down with his ship.

United States Navy Regulations instruct the captain that 'he shall, in the case of loss of the ship, remain by her with officers and crew as long as necessary, and save as much Government property as possible' which is little different from the British regulation. The US Navy Regulations then go one step further by stating that 'If it becomes necessary to abandon the ship, he [the captain] should be the last person to leave her.'[4] This is, however, very far from suggesting that the commanding officer should not leave his ship at all.

The concept of the 'captain going down with his ship' has long been accepted not only in naval circles but also in certain civilian ones, and references in newspapers to captains doing this are almost invariably couched in laudatory – or, at the very least, non-condemnatory – terms. On 22 June 1893 the British Mediterranean Fleet was preparing to anchor off the Lebanese port of Tripoli (now Trablous) when, owing to confused orders, two battleships collided while attempting to execute a turn, HMS *Camperdown* hitting the flagship, HMS *Victoria*. Many died, including the commander-in-chief Admiral George Tryon, but *Victoria*'s commanding officer, Captain Maurice Bourke, survived. Back in London, the wife of Captain Gerard Noël, who had been commanding HMS *Nile* and had witnessed the disaster, wrote to her husband telling him that 'I do wish Captain Bourke had stuck to his ship. He will never have such prestige again. Never. Everyone says the same.'[5] Thus, Mrs Noël and her London friends were not only aware of the practice but were also clearly in favour of it.

Nor has the concept been confined to European and North American navies, as witnessed by events in Buenos Aires in December 1939. As described in Chapter 9, having blown up his ship, the *Admiral Graf Spee*, in the River Plate estuary on the evening of Sunday 17 December, Captain Langsdorff and his crew were transported by boat to Buenos Aires. There, the following morning, the Argentine newspapers castigated the

German commanding officer 'as a coward and a traitor to the tradition of the sea *because he had not gone down with his ship*' [my italics].[6]

Furthermore, the concept is still alive, as evidenced by a 1996 official reappraisal of the loss in 1945 of USS *Indianapolis* (CA-35), commanded by Captain Charles McVay III. In a top-secret operation, *Indianapolis* delivered crucial parts of the two atomic bombs to be used against Hiroshima and Nagasaki to the island of Tinian, and then set off for Leyte in the Philippines on the return voyage. McVay's orders were to zigzag if a submarine threat existed, but by that stage in the war very few Japanese submarines remained operational and the Philippine Sea was considered to be safe. Nevertheless, on 29 July McVay zigzagged during daylight hours, reverting to a straight course only with the approach of darkness. That evening, however, *Indianapolis* chanced upon the Japanese submarine *I8* (Captain Hashimoto Mochitsura), which launched six torpedoes, scoring hits with four. *Indianapolis*, with 1,119 aboard, took just twelve minutes to sink. Though it is estimated that 800 survived the sinking, owing to a variety of errors and misunderstandings they were not found for five days, by which time only 320 remained alive. The sinking was a tragedy for which McVay, one of the survivors, took immediate and full responsibility, but, unusually for the US Navy, he was court-martialled for losing his ship. He was found guilty, although the sentence was the most lenient available to the court, involving only a loss of seniority.[7]

The way in which McVay had been treated caused considerable ill-will among sections of the US Navy and the public, a feeling that was heightened when in 1966 McVay, by then a rear-admiral on the retired list, donned his uniform and committed suicide. The controversy continued and an official review was conducted in 1996 by a naval lawyer, Commander R.D. Scott. He commented that 'The Navy has never challenged Captain McVay's uncorroborated account that he did not *go down with his ship* [my italics] because he was swept over the side by a wave.' This is a most unusual allegation, not least because it was made by a serving naval lawyer in an official document and contains a very clear implication that McVay (and presumably other captains in similar situations) had some obligation to 'go down with the ship'.

Sometimes things do not turn out quite as the captain had intended. Following the battle of Tsushima in 1905, the captain of the Russian battleship *Admiral Nakhimov* went down with his ship but floated back to the surface and was found unconscious but still alive by some Japanese fishermen. Similarly, Commander Usher of HMS *Valerian* also had every intention of going down with his ship when it foundered in the Caribbean in 1926, but he was hit on the head by a falling spar, as a result of which he too floated back to the surface still unconscious and was pulled on to a raft and rescued. Aware of such a danger, on several occasions in the Second World War Japanese captains had themselves tied to the binnacle by their crewmen just before the ship went down, thus ensuring that they did not accidentally survive.

Quite why some captains consider it necessary to go down with their ships and others do not is unclear and certainly does not seem to have depended on the speed at which the ship has sunk. During the Second World War, Captain Makeig-Jones went down with HMS *Courageous* when she sank on 17 September 1939, as did Captain D'Oyley-Hughes of HMS *Glorious* on 8 June 1940. When another carrier, HMS *Ark Royal*, was torpedoed on 13 November 1940, her captain and crew fought for a day to save her. However, when it was eventually realized that her loss was inevitable she was abandoned in good order, and her commanding officer, Captain Maund, was the last to leave. There has never been any suggestion that Maund should have remained with his ship. Similarly, when the destroyer HMS *Kelly* was sunk in action on 23 May 1941, her commanding officer, Captain Lord Louis Mountbatten, survived, and nobody – as far as is known – thought any the worse of him for having done so.

The decision to go down with the ship is made in unpre-meditated circumstances and, at the risk of stating the obvious, the motives of those who have done so can only be guessed at. In some cases the commanding officer may have felt a sense of failure at losing his ship, and the possibility of court martial and disgrace may have had some bearing on his decision. Yet there are others – for example, Glasford of the *Acasta* – who clearly did far more than could have been expected of them but still chose to follow this course. Whatever the case, it must take great personal courage for a commanding officer to force himself to remain aboard his ship as it sinks, when all his senses must be urging him to attempt to escape.

In the strictest sense, it could be argued that 'going down with the ship' is, in some respects at least, an abandonment of responsibility. Most immediately, the absence of the captain makes the next senior surviving officer responsible for assem-bling the survivors and for doing his best to ensure that they remain alive until rescued. Once rescued, the same officer must then explain the circumstances of the loss to the naval author-ities, even though he may not have been in the captain's confi-dence at the time. Indeed, in many navies it is the practice that the captain and his second-in-command should be in physi-cally separate parts of the ship to ensure that one explosion cannot kill them both.

Furthermore, the captain is, virtually by definition, a valued and highly trained officer. By going down with his ship he deprives his country of his services for the remainder of the war.

While the origin of this practice and its perpetuation over a period of 150 years remain a mystery, there can be no doubt that, in the vast majority of cases, a captain who 'went down with his ship' was demonstrating both his personal courage and his overwhelming sense of commitment to his ship. Whether he was wise to do so, or acting in the long-term inter-est of his service, is quite a different matter. Interestingly, there have been no known cases of a captain taking this course since the end of the Second World War, and it is to be presumed that this unusual practice has now passed into history.[8] Certainly, there are other – and much better – ways in which a captain can demonstrate his commitment to his ship and his crew.

11

A Sort of Peace

The second half of the twentieth century saw numerous military interventions in foreign countries in conditions that fell short of war. The role of those involved ranged from peacekeeping, peacemaking and humanitarian operations to reprisals and the rescue of hostages. All these missions required the traditional attributes of a commanding officer, but some also demanded new qualities of diplomacy and tact, coupled with the ability to operate in an environment where the military force was strictly neutral and could resort to combat only as a final option. In addition, such missions were sometimes organized along unconventional lines and involved a complex chain of command. In these circumstances the commanding officer was not only in the front line of military operations, which was only to be expected, but was also at the centre of political and media attention.

Although foreign military intervention in domestic crises is often seen as a post-Second World War phenomenon, it is in fact nothing new. In December 1861 a joint force of British, French and Spanish troops occupied Vera Cruz in an effort to compel Mexico to meets its debt obligations, while another international force, this time a combined Anglo-German naval squadron, carried out a similar operation against Venezuela in 1902–3. The United States also intervened regularly in Central America. Vera Cruz was occupied for eight months in 1914 as a reprisal for insults to US sailors in Tampico, while longer occupations took place in Dominica from 1916 to 1924 and in Haiti from 1913 to 1934.

In the late twentieth century, two new factors contributed to the increase in military interventions. First, the ever-growing availability and capability of air transport meant that larger and more militarily capable units could be dispatched to trouble spots with ever greater rapidity. This meant that units could often deploy so quickly that they would outstrip both the higher command's ability to establish a command and control system for the operation and the intelligence system's ability to brief them on local conditions. Second, the 'international community' increasingly formed multinational forces to attempt to reduce the impact of civil war, notoriously one of the most vicious types of conflict.

LIEUTENANT-COLONEL AUSTIN

Koh Tang Island, 15 May 1975

On 30 April 1975, as helicopters lifted the last refugees out of Saigon, the long, painful and costly American involvement in Indochina appeared to be at an end. Then, just two weeks later, at about 1130 on 12 May 1975, a group of armed Cambodians hijacked the United States container ship *Mayaguez* and her forty-man crew in international waters. The ship's distress call was picked up and the information reached Washington at 1612 that afternoon, although by then the master had been forced to take his ship to an anchorage one mile north of the island of Koh Tang, anchoring there at midday on 13 May. Thus, without warning, the American government found itself faced with another major problem in south-east Asia. Since the United States had no diplomatic relations with Cambodia and since American lives were apparently at risk, there appeared to be no alternative to military action. Accordingly, President Ford ordered an immediate rescue operation.

The commander of the Seventh US Air Force, located at Nakhon Phanom in north-east Thailand, was designated as the 'on-scene operational commander' and he immediately deployed a staff element and all available helicopters to the nearest US base, at Utapao on the Thai coast. The only ground troops immediately available were 125 airforce policemen from Nakhon Phanom's base security force, and the first plan, conceived in great haste, was to send these by helicopter to retake the *Mayaguez*. They were, however, totally unprepared for such a role, and it was quickly (and very fortunately) decided to use Marines instead.

Task Group 79.9 was created for this mission with a Marine Corps officer, Colonel Johnson, designated as its commander. Accompanied by four staff officers, Johnson arrived at Utapao at 0730 on 14 May. There he was told that his mission was to retake the *Mayaguez*, recover her crew and seize, occupy and defend Koh Tang. To achieve this he was allocated two combat elements: Battalion Landing Team 2/9 (BLT 2/9) (based on the 2nd Battalion of the 9th Marines) commanded by Lieutenant-Colonel Austin, and Company D of the 1st Battalion of the 4th Marines (D/1/4), commanded by Major Porter. Meanwhile, the Seventh Fleet started to move ships into the area, the nearest being the aircraft-carrier USS *Coral Sea* and the destroyer-escort USS *Holt*. A second aircraft-carrier, USS *Hancock*, and other ships sailed from Subic Bay in the Philippines.

Porter's D/1/4 was based in the Philippines but moved quickly, flying into Utapao on 0505 on 14 May. Once there, it became the victim of military confusion. Initially it was briefed for a heliborne assault on the *Mayaguez* and put on 30 minutes' notice to move. Then, at 1200, it was ordered to board the helicopters. At 1400 it was told to disembark and revert to 30 minutes' stand-by, remaining at that status until 0200 the following morning. The other ground unit, 2/9 Marines, was training on Okinawa when Austin received the order to deploy at 2030 on 13 May. He quickly disengaged his unit from the exercise, took them back to their base and then moved to the nearest airport. Austin arrived at Utapao on 0945 on 14 May, where he, Porter and Johnson got down to planning possible operations.

Koh Tang is a low-lying, jungle-covered island. Its northern tip is connected to the rest of the island by a neck. On either side of the neck lie two beaches linked on either side by a cut 55 yards wide. First light was at 0542 and an early morning haze cleared by about 0900, after which the temperature climbed steadily throughout the morning, reaching the mid-90s by midday. Humidity was high. Sunset was at about 1730 and, as usual in the tropics, was both sudden and short-lived. On the day of this operation there was no moon, resulting in an almost total blackout.

Austin was allowed one aerial reconnaissance in a light transport aircraft, the 400-mile round trip lasting two and a half hours. The pilot had orders to remain above 6,000 feet to avoid alerting the enemy. This meant that the observers were unable to see the ground in any detail, although Austin did spot the beaches on either side of the neck and the cut connecting them, and was able to observe that these were the only feasible helicopter landing sites. He also saw the *Mayaguez* lying at anchor to the north of the island. The only other information came from two Cambodians who told Austin that the garrison numbered about 20 or 30 lightly armed irregulars.

Austin's knowledge of the island and its defenders was therefore minimal. He had no map, he was given only one poor-quality aerial picture (which was supposed to suffice for the entire battalion), he had been allowed only one aerial reconnaissance of limited value, and he had been grossly misinformed (although he did not know it at the time) by the two local 'experts'. On top of that nobody could tell him where the crew of the *Mayaguez* actually was.

The helicopters assigned to the mission, all from the United States Air Force, were Sikorsky H-53s but of two sub-types which had minor but significant differences (see table 13). The two types were known by their radio callsigns as 'Jolly Green' and 'Knife' respectively. Austin was allocated eight helicopters

TABLE 13: *Helicopter Characteristics*

	Jolly Green (JG)	Knife (K)
Nickname/radio callsign	Jolly Green (JG)	Knife (K)
Official designation	HH-53B	CH-53C
Primary role	Search-and-Rescue	Special Operations
Crew	6	4
Passengers		
Normal	38	38
This operation	20	27
Air-to-air refuelling	Yes	No
Additional armour	Yes	Yes
Cruising speed (typical)	173m.p.h.	
Range, loaded (typical)	540 naut. miles	
Weapons (typical)	3 x 7.62mm Miniguns	

for his part of the plan: three HH-53s (Jolly Greens) and five CH-53s (Knives). That meant he could transport 195 men. The flight time to Koh Tang was estimated at 1 hour 45 minutes, in other words a round trip of approximately four hours.

BLT 2/9 consisted of the 2nd Battalion 9th Marines (four companies, each of four platoons, plus a headquarters company), I Battery 3/12 Marines (Artillery), and a platoon each of engineers, 81mm mortars and 106mm recoil-less rifles. The mission changed several times, but the team was finally ordered 'to seize, occupy and defend the island of Koh Tang, hold the island for a minimum of forty-eight hours and to rescue any of the *Mayaguez* crewmen who are assumed to be on the island'. Company D/1/4's mission was 'to seize the *Mayaguez* and remove it from its current location'. The only firm timings Austin was given were that take-off would be at 0415 and that the arrival of BLT 2/9 on Koh Tang was to coincide with that of D/1/4 aboard the *Mayaguez*. Austin's plan was for Company G to form the first wave, reinforced by a section of 81mm mortars, and accompanied by a small command element, led by himself and including a doctor, linguists, bomb disposal experts and other specialists intended to look after the *Mayaguez* crew. Two helicopters were to land on the western beach, the remaining six on the eastern beach.

However, given the uncertainty over the whereabouts of the *Mayaguez*'s crew, no preparatory fire on the landing zones was permitted. Company E was to form the second wave, with Companies F and H, and I Battery 3/12 Marines, held in reserve. Austin issued his orders at 2200 and after only a short rest the force paraded at 0230 on 15 May. The first helicopters lifted off at 0415, as planned.

The first two helicopters approached West Beach at dawn, Knife 21 (K21) leading, and all was quiet until the last Marine left the aircraft. The enemy then opened fire from concealed positions at short range, using rifles, rockets and mortars. Surprise was total and K21 was badly damaged, the pilot being able to fly only a short distance before ditching in the sea. The second helicopter, K22, tried to land its Marines in support of those already ashore, but was repeatedly hit. Forced to abort, it managed to fly to the mainland where it made an emergency landing on a beach. K22's passengers included the assault company commander, who was thus temporarily removed from the scene of action, but his executive officer was already ashore and took command.

The next aircraft, K32, dumped fuel in order to rescue K21's three surviving crewmen, then headed for the beach to land its Marines, but it incurred so many hits that it too had to make an emergency landing on the mainland. The other aircraft in this wave, Jolly Green 41 (JG41), made several attempts to land its Marines but was driven off by fire, following which, short of fuel, it was forced to leave for air-to-air refuelling with a nearby Hercules HC-130P tanker.

The first assault landing on East Beach fared even worse, with K23 and K31 both coming under heavy fire as they were about to touch down. K23 was hit immediately, forcing the pilot to slam it on to the beach where he ordered the passengers and crew to abandon it. Surprisingly, only one person was injured, a flight mechanic who received a head wound while dashing up the beach. K31 was hit by a rocket grenade and burst into flames. It then fell some 25 feet into the water, where twelve men were killed. The survivors swam out to sea, where they were subsequently rescued, but it was thought that one wounded man had taken shelter in the wreck of K23.

Meanwhile, Porter's D/1/4 had successfully boarded the *Mayaguez*, which was found to be deserted, and she was then towed out of the area by USS *Holt*. Thus, in the first hour of the mission, while the Americans had regained the ship, there were still twenty-five men ashore on East Beach, twenty-nine on West Beach and thirteen swimming in the sea. Three helicopters had been lost, two had made forced landings on the mainland and one was refuelling. It was hardly an auspicious start.

Three more helicopters now arrived off West Beach – JG42 and JG43 making their first approach, and JG41 returning after refuelling – but all three were initially driven off by heavy fire. JG42 eventually managed to land its passengers but in doing so incurred so much damage that it was forced to return to the mainland. JG41, after further unsuccessful attempts to land, withdrew to refuel for a second time.

Austin was aboard JG43. As they approached he saw a column of smoke which he assumed to be the result of a US air attack. He was quickly corrected by a message from the pilot that two helicopters had been downed on East Beach, his first intimation that all was not well. Then his helicopter was unable to land on West Beach and the pilot flew along the coast until he found a small spot 1,300 yards further south. At about 0615, the passengers landed, following which JG43 returned northwards to escort the damaged JG42 back to the mainland.

At about 0800 hours the three helicopters which had taken part in the retaking of the *Mayaguez* became available: JG11 and JG21 went to Utapao to collect more Marines, while JG13 flew to Koh Tang to try to extract the twenty-five men pinned down on East Beach. The pilot managed to land but intense enemy fire not only prevented the Marines from leaving their cover on the jungle edge but also damaged the aircraft so badly that the pilot had to lift off again and return to the mainland for repairs. During this particular episode, however, a recently arrived airborne forward air controller managed to pinpoint

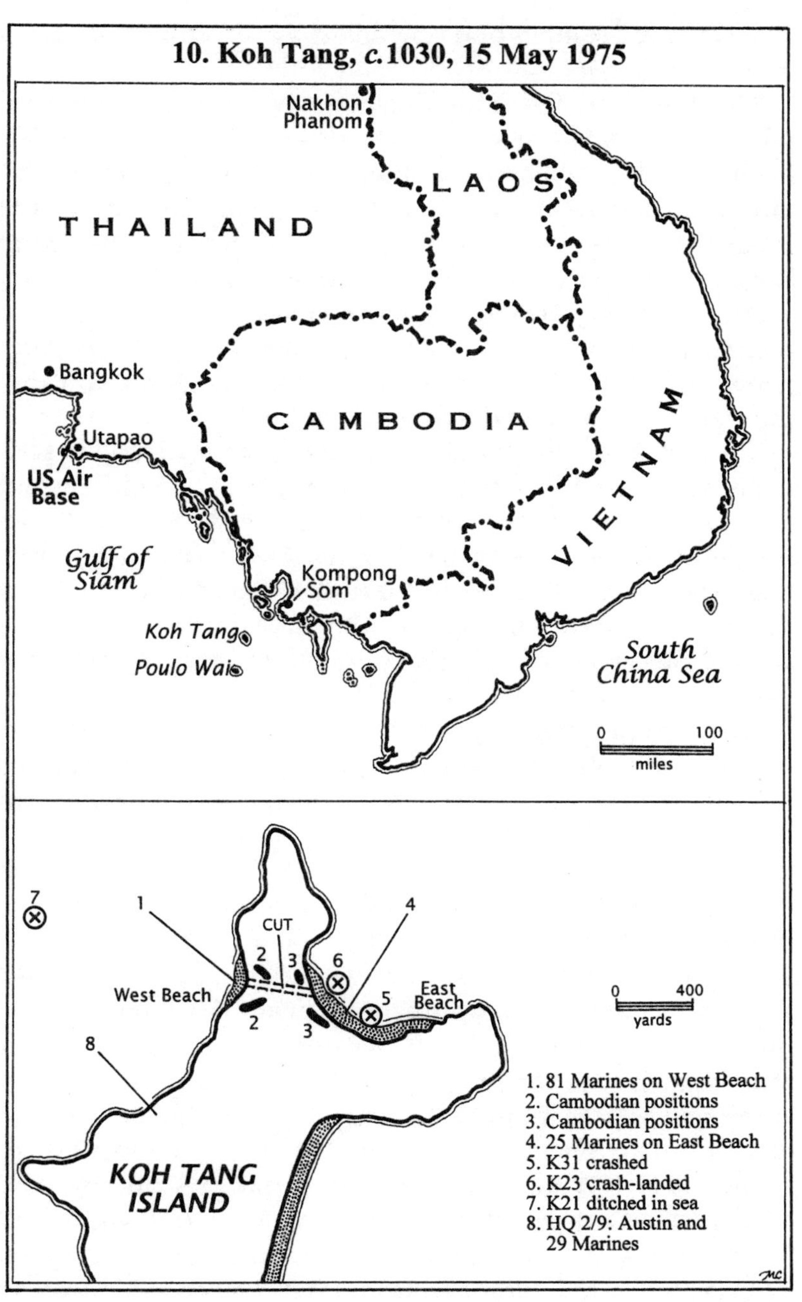

10. Koh Tang, *c.*1030, 15 May 1975

some enemy positions and called in close air support to attack them.

JG41 then returned from refuelling and this time, with supporting fire from a C-130 'Spectre' gunship, the pilot succeeded in landing. However, his machine was seriously damaged by mortar fire and he was forced to take off with five Marines still aboard. After air-to-fuel refuelling a third time he returned to base. He and his crew had been airborne by eight hours. By about 1100 fifteen Americans had been killed and many more wounded, twenty-five were still trapped on East Beach, and twenty-nine were fighting their way up the west coast towards the eighty-two pinned down on West Beach.

It was at this point that the entire crew of the *Mayaguez* suddenly reappeared. They had initially been held on Koh Tang but had then been moved to another island just off the coast. When first sighted, they were aboard a local fishing boat flying a white flag. They were quickly transferred to the recently arrived destroyer, USS *Wilson*, which took them to the *Mayaguez*. Once aboard they soon got under way and the *Mayaguez* continued her interrupted voyage.

This development changed the situation dramatically. Higher command now tried to prevent the five helicopters then approaching Koh Tang with marine reinforcements from delivering them. On the ground and deep in battle, Austin was never officially told of the return of the hijacked crew, but on hearing that the reinforcements had been turned back he insisted that they be rerouted to Koh Tang.

The first aircraft of the second wave, K52, approached West Beach at about 1230 but came under heavy fire which caused leaks in the fuel tanks. Since it was now incapable of air-to-air refuelling, it was forced to make for the mainland without landing. The remaining four aircraft, however, managed to deliver their troops who included the commander of Company G. K51 and JG12 also managed to pick up a number of casualties and take them back to the mainland. During all this helicopter activity on West Beach (and perhaps because of it), Austin and his command team finally managed to join up with the men in the main position, having fought their way through several enemy positions and captured a 60mm mortar, a 90mm rocket-launcher and numerous rifles on the way.

At this point, there were 222 men ashore, with 25 pinned down on East Beach and 197 on West Beach. The latter were more secure and had managed to penetrate some distance inland towards East Beach, but they were being held up by a Cambodian position designated 'Camp'. It was also at this time that Austin, now on West Beach and catching up with events, was told by the newly arrived commander of Company E that the crew of the *Mayaguez* had already been freed. In the circumstances, his original mission to seize and hold the island for at least forty-eight hours no longer applied, but with his Marines confined to two pockets and with the next helicopter wave four hours away he decided that he had no option but to consolidate and secure his position.

Another attempt to extract the men from East Beach by helicopter was made at about 1430. The enemy positions were bombarded by close-support aircraft and naval guns (several destroyers were now off shore). JG43 then attempted to land but was seriously damaged and had to withdraw, following which it flew to the carrier *Coral Sea*, now only about 70 miles away, for repairs.

During JG43's attempted landing one of the Marines ashore spotted fire coming from a local boat grounded in the bay. Up until then it had been assumed to be abandoned, but now it became evident that it was manned by Cambodians who were firing at any helicopter within range. As a result, air support was called in and the men in the boat ceased to be a problem. JG11 was then able to reach the beach under cover of supporting fire and, although the helicopter still came under intense fire from other positions, it managed to recover all twenty-five men and deliver them to the approaching *Coral Sea*. It was thought, however, that there was still one wounded American in the wreckage of K23, so JG12 spent several minutes hovering above it while the winch man inspected the wreck. No sign of life was found but JG12's winch operator was wounded and the aircraft damaged, so it too flew on to the *Coral Sea*. East Beach was now clear.

Following some hasty repairs to its fuel pipes, JG43 now returned from *Coral Sea*, while JG44 made a first appearance at Koh Tang, having taken part in another, unrelated operation. These two, plus the still undamaged K51, were all that were left to rescue the marines still on the island, although two destroyers (*Holt* and *Wilson*) and the aircraft-carrier *Coral Sea* were now close by. Also at about this time a C-130 Hercules arrived high over the island, dropped an enormous 15,000lb blockbuster bomb and departed. The explosion blasted a huge hole in the jungle, but who authorized the mission, why, and what it was supposed to achieve were never discovered – and the Marines on the ground had absolutely no warning of its arrival.

Although Austin was nominally under the command of the Task Group's Commander, Johnson, at Utapao, the two Marines had no direct communication with each other following Austin's departure from Utapao. Once on Koh Tang, Austin's only communications rearwards were to an Airborne Battlefield Command and Control Center (ABCCC), a modified C-130 orbiting 90 miles away, which carried the Airborne Mission Commander (a USAF colonel) and his staff. These men controlled the tactical aircraft operating in the area of the island, but as far as Austin was concerned the ABCCC's main purpose was to relay messages orally between himself and a mysterious 'they', whose identity at the time remained unknown to Austin. In fact, 'they' were HQ Seventh Air Force, which was effectively in direct command of the operation.

To add to the communication problems on the island, most of those involved in the operation simply latched on to Austin's battalion command radio net. In addition to its many normal users, it therefore became swamped with communications from helicopters, fixed-wing aircraft, naval ships, airborne forward air controllers and the ABCCC.

In the course of the afternoon the ABCCC relayed a series of questions to Austin which led him to believe that the higher command was planning an extraction operation. He provided the answers (Could small boats from USN ships be used? Yes. Could extraction be done entirely from East Beach? No. Could extraction be done at night? Probably.), but he also added the important proviso that once an extraction began it must be pursued until complete, since a reduced force could not be left where it was. No word on any final decision was ever received, and the first Austin knew that extraction was taking place came after dark, when three helicopters were heard approaching.

First in was K51, guided by Marines shining flashlights. Despite the now customary hostile small arms fire it managed to land and lifted off with a full load of Marines. It was followed by JG43 and then JG44, both of which recovered a full load, leaving seventy-three Marines in a small perimeter on the jungle edge and under intense fire. Sensing the urgency of the situation, the pilot of JG44, Lieutenant Blough, instead of heading for the *Coral Sea*, flew to the much closer *Holt*, thus saving twenty minutes' flying time. Transferring men from a large helicopter to such a small destroyer-escort would have been extremely hazardous in daylight, but at night and without lights (JG41's landing lights had been shot out) it could reasonably have been considered impossible. Fortunately, nobody told Blough, and with his crewman hanging out of the door and giving directions he managed to transfer his passengers to *Holt* and then return for another load. This time, however, engine problems forced him to return to the *Coral Sea*. That left twenty-nine men ashore in a most precarious situation, but fortunately K51 returned and managed to land on the beach at the third attempt and extract them all. Austin then found that his men were scattered – three groups were aboard *Coral Sea*, *Holt* and *Wilson* at sea, now a fourth group was ashore on the mainland. It took him nearly two days to establish that three men could not be (and never have been) accounted for.

The action at Koh Tang lasted fourteen hours, during which the Americans lost thirteen killed (all in the first ninety minutes), three missing presumed dead, and forty-three wounded, while three helicopters were total losses and eight were seriously damaged. Cambodian losses were never known but must have been substantial.

In all events such as the *Mayaguez* incident, the government concerned is faced, first of all, with the need to respond rapidly to prevent the incident escalating, to save lives and to demonstrate determination. Against this, however, must be balanced the requirement to gather intelligence and to select and rehearse the rescue force in order to launch the rescue with the maximum likelihood of success. All this must take place against the background of a vociferous media and, usually, of increasing political and public pressure for 'something to be done'. Thus, in the case of the *Mayaguez*, there was very strong pressure from within the United States, from President Ford down, for rapid and determined action. However, as is normally the case, the man who actually bore the responsibility was the commander on the ground, in this case, Lieutenant-Colonel Austin.

On arrival at Utapao, Austin had found himself faced with a situation in which there were no maps and only one poor-quality air photograph. Although the air reconnaissance undertaken was of some value, it took up a large proportion of the available time, while the minimum height restriction, which was imposed for security reasons, meant that he saw comparatively little. Then the intelligence briefings were both inadequate and in many instances inaccurate. Ironically, a few days after the operation, an intelligence report surfaced which was dated two days *before* the operation and which gave an almost entirely accurate picture of the Cambodian force – 150 men armed with automatic rifles, light machine-guns, 60mm mortars and 90mm recoil-less rifles.

There were other factors over which Austin had no control. While plenty of close air support became available later in the day, the lack of preparatory strikes before the initial landings made the large, slow-moving H-53s very vulnerable. In addition, the distance between the base at Utapao and Koh Tang meant that the second wave arrived four hours after the first, resulting in an almost fatally slow build-up.

The command system, cobbled together at short notice, was inadequate. Despite being far from the scene and not in direct touch with Austin, the commander of the Seventh Air Force appears to have exercised close direction of events through the ABCCC; indeed, as far as the Marines were concerned, the latter's only significant function was to provide a relay to the Seventh Air Force. Nevertheless, despite the existence of this link Austin was repeatedly starved of information vital to his conduct of the operation on the ground, while Colonel Johnson, who was nominally in tactical command of 2/9BLT and Company D 1/4 Marines, was placed in a remote corner of Utapao airfield with such poor communications that he had no direct link with either Austin or Porter.

Despite all these shortcomings the Marines successfully undertook a rapid deployment to Utapao. Co-operation between the USAF helicopters and the Marines, who had not worked together before, was effective, as was close air support over the island. On the ground, Austin and his Marines remained remarkably cool in a confusing and highly dangerous situation.

With hindsight, however, many things could have been done better. In particular a delay until the following day would have enabled the two aircraft-carriers to be used to mount the operation from close offshore (indeed, the operation might not have been required at all since the *Mayaguez* and its crew could by then have been released). This would also have given more time for the gathering of accurate information and for the establishment of a tauter command chain.[1]

LIEUTENANT-COLONEL BOB STEWART

UN Protection Duties in Bosnia, 1992–3

Unlike other examples in this book, the tour of duty of the 1 Cheshire Battalion Group in Bosnia did not include a specific action in which the military skills and capabilities of the commander and his unit were put to the test. Instead, for a period of six months, they had no idea what challenge each new day would bring. Because it was so unusual and crucial to all that happened, it is necessary first to take a brief look at the political environment in which Colonel Stewart had to operate.

The history of the Balkans is both long and complicated, but a brief resumé of what had occurred since the Second World War will suffice here. In 1945 Marshal Tito established six republics within Yugoslavia – Bosnia-Hercegovina, Croatia, Macedonia, Montenegro, Serbia and Slovenia – and two auton-omous zones, Kosovo and Vojvodina. For the next thirty-four years he held the country together by a mixture of strength of personality, a powerful and often oppressive state machine, and judicious distribution of power between the three major groups, Croats, Muslims and Serbs.

Following Tito's death in 1980, Serbia tried to perpetuate Yugoslavia, but with the aim of establishing a Serb-dominated, pan-Balkan state. To this end it fought first Slovenia and then Croatia in 1990–2 to prevent their secession. This led to the initial deployment of the United Nations Protection Force in four areas of Croatia in January 1992. Eventually, both Croatia and Slovenia managed to attain independence, and they were followed, albeit more peacefully, by Macedonia.

That left Bosnia-Hercegovina (hereafter Bosnia) as the major subsidiary element of the Serb-led republic. Bosnia had a pop-ulation of 5 million, in which no major group formed an abso-lute majority, 43 per cent being Muslim, 33 per cent Serbs (Orthodox Christians), and 17 per cent Croats. In March 1992 the Muslim-led government of Bosnia obtained a large major-ity in a referendum on independence, following which, with US encouragement, it broke away from the former Yugoslavia (effectively, Serbia and Montenegro) on 5 April 1992. A violent civil war within the newly independent country then immedi-ately broke out. On one side was ranged the Bosnian Serb army, 70,000 strong, generally well-trained and well-armed, and with strong support from the Serbian army. On the other side was an uneasy alliance of the (Bosnian) Croatian Defence Council,

45,000 strong, with direct support from the Croatian army, and the Bosnian Muslim army, which numbered 50,000 but which was less well equipped, organized and trained than the other two and which had no outside support. The professional core of all three armies in this Bosnian war was provided by officers of the former Yugoslav National Army. These men knew each other and were trained in essentially the same tactics and the use of the same equipment. All three Bosnian armies were also supported by large numbers of paramilitaries.

The aim of the Bosnian Serbs, located in western and eastern Bosnia, was to join these two territories and to expel all non-Serb inhabitants. By mid-1992 they were shelling Sarajevo and putting pressure on other areas, so that an increasing number of refugees were forced to flee their homes. This process, which came to be known as 'ethnic cleansing', was graphically depicted on global television.

Outside powers, particularly France, Germany, Russia, the United Kingdom and the United States, and the European Union and the United Nations, tried to find a solution to the conflict. Their efforts were, however, constantly thwarted by one or other of the parties. Nevertheless, increasing domestic pressure in France, Britain and the United States required 'something to be done' and, in the absence of anything more positive, this led to a commitment to provide military protection for the distribution of humanitarian aid, a novel form of involvement by armed forces.

The major participants were France and Britain. In August 1992, the British government announced that it was considering contributing a battalion group to a new segment of the United Nations Protection Force, located in the Balkans, and that if it did, the group would be based on the 1st Battalion, Cheshire Regiment (the Cheshires).[2] The commanding officer of this British military contingent was Lieutenant-Colonel Bob Stewart, aged 43. Commissioned in 1969, he had undergone several tours in Northern Ireland, obtained a degree in international politics, attended the Army Staff College and held several high-grade staff appointments. He was appointed commanding officer in March 1991 and took the battalion to Fallingbostel, Germany, in November 1991. There he trained his unit for conventional armoured-infantry operations, although one of his main tasks was to run down numbers before a planned amalgamation with the Staffordshire Regiment in 1993.

After several reconnaissances in September and October 1992, the battalion's advance party arrived in Split on 29 October. The main body began to arrive at the operational base in Vitez on 10 November and was declared operational under the command of the United Nations Bosnia-Hercegovina Command (BHC) on 17 November. Stewart's command comprised the Cheshires, brought up to full war strength by one hundred men from the 2nd Battalion of the Royal Irish Regiment, plus a Medium Reconnaissance Squadron of the 9/12th Royal Lancers and 43 Field Squadron of the Royal Engineers. The battalion was responsible for its own internal administration, run by an organization known as 'Echelon', and had immediate medical support at its base from a field surgical team. The battalion group also included a 'public information' (PInfo) team, plus fifteen locally employed interpreters and, later, six UK liaison officer teams.

A rear party remained in Germany and was responsible for the administration of the peacetime base and, in particular, for looking after the families of those troops serving in Bosnia. Administrative support of the battalion group was a British national responsibility, with two HQs in the area, HQ British Forces at Split, and HQ UK National Support Element at Tomaslavgrad.

The battalion group took its full war equipment with it, including support weapons such as mortars and machine-guns. The majority of the infantry were mounted in Warrior tracked combat vehicles, but a number had older tracked armoured personnel carriers and wheeled Land Rovers, while the Reconnaissance Platoon was mounted in Scimitar tracked vehicles, as was the 9/12L squadron.

The Cheshires' base comprised two sites on the Vitez-

Travnik road. The main site was an abandoned secondary school two miles from Vitez, which initially accommodated Battalion HQ, three company groups, and the cavalry and engineer squadron HQs, while 'Echelon' was approximately a mile away in an abandoned garage. Since few buildings remained, most troops were accommodated in tents, but from late December onwards steps were taken to increase protection, particularly against incoming artillery shells.

Throughout the period one company group was based at Gornji Vakuf with the task of keeping the route from the coast to Vitez open and secure. This company originally consisted of three platoons mounted in fourteen Warriors, plus the Cheshires' Reconnaissance Platoon (eight Scimitars), but from late November onwards one platoon was withdrawn to Vitez to provide the stand-by platoon.

One of Stewart's early priorities was to get aid through to Tuzla. He began by opening a forward operating base at Kladanj in late November, but in late December he was able to push this forward to Tuzla itself, closing the Kladanj site. In Tuzla a company group, with three Warrior platoons and a Scimitar troop, used existing buildings on a deserted airfield.

Since it was important to rotate both sub-units and individuals, the platoon detached from Gornji Vakuf to Vitez and the company at Tuzla were both changed over at six-week intervals, while from early December onwards all men were sent on compulsory 'rest-and-recuperation' (R&R) leave to either Germany or Britain. This deprived the battalion group of seventy men at a time (approximately 10 per cent of its strength), but the resulting benefits in terms of morale far outweighed the tactical disadvantages.

The engineers (130 strong) were spread thinly, with some men in two 'mountain camps' whose task was to keep open the stretch of the supply route running through the mountains between Tomaslavgrad and Gornji Vakuf. Other engineers were involved in tasks such as bridge-building, mine clearance and constructing and improving facilities at the various bases.

As in any international operation, Stewart had to answer to

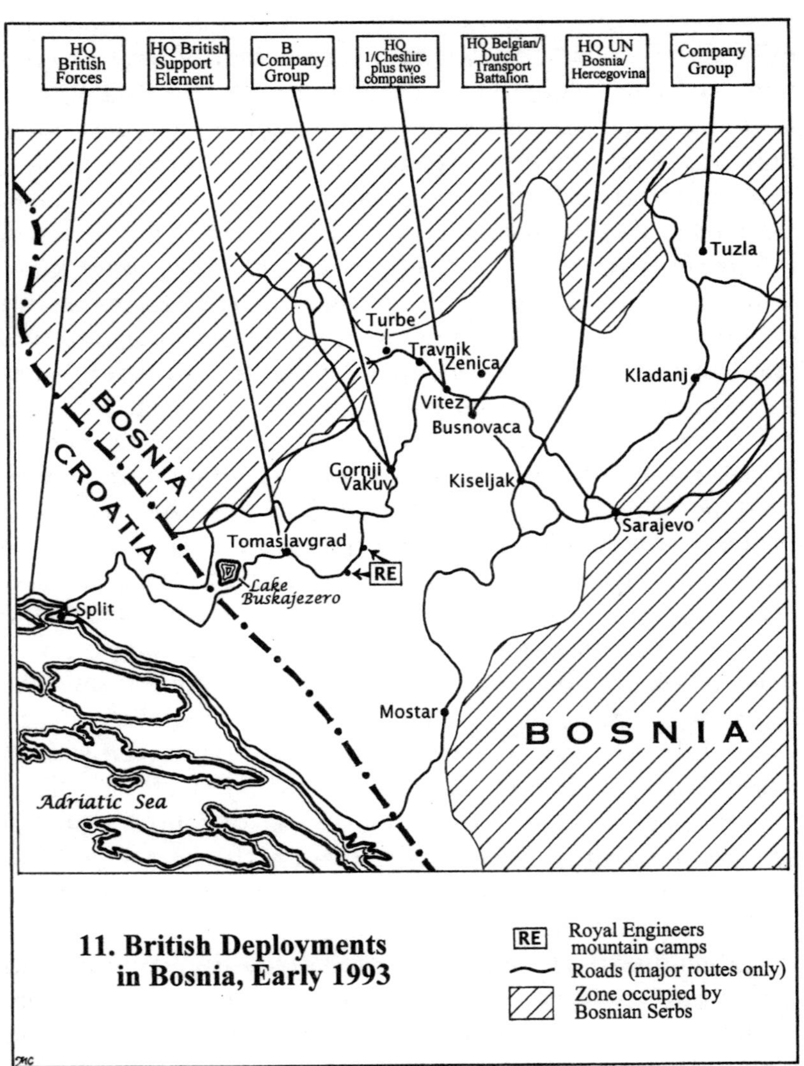

11. British Deployments in Bosnia, Early 1993

two chains of command. The national, that is British, author-ities retained some control over his activities, his immediate national superior being the Commander British Forces (Com-britfor), a brigadier-general located at Split, who commanded all British forces in Bosnia and Croatia, and whose national superior was HQ UK Land Forces at Wilton in England. Administrative support was also a national responsibility, being conducted by the British National Support Element located at Tomaslavgrad, with forward detachments at Vitez and Gornji Vakuf.

Stewart's operational chain of command was to the United Nations BHC, commanded by the French Major-General Philippe Morillon, whose HQ was at Kiseljak outside Sarajevo. This came under the Commander UN Forces who, for most of Stewart's tour, was General Nambiar of the Indian Army.

As with the chain of command, so communications respon-sibilities were split between the United Nations and Britain. The Cheshires' communications with the UN were provided by a Dutch signal battalion, and the Dutch also manned the communications centre located at Stewart's main base at Vitez. These communications were never particularly good and were always at their worst at night, which was of particular concern when trying to talk to HQ BHC at Kiseljak. Communications within the battalion group and with Britfor at Split and HQ Land Forces in England were a British responsibility and worked well.

Civil communications were, not surprisingly, much affected by the conflict. Although the telephone system was still in place, it operated only intermittently, while a mobile telephone system, also in place, had only limited coverage. One feature of both systems was that when they did work, anyone could speak to anyone else, regardless of which side they were on. All forces involved in the Bosnian civil war used radios and radio procedures from the former Yugoslav army, and the opposing sides could communicate with each other simply by finding a frequency being used by the other side, breaking in, passing a message and then awaiting developments.

The United Nations Protection Force (Unprofor), around 7,000 strong, had two formal missions: to support the UN High Commissioner for Refugees (UNHCR) in the distribution of humanitarian aid throughout central Bosnia; and to escort detainees to a place of safety after their release had been arranged by the International Committee of the Red Cross. A vital factor in this was that Unprofor was deployed neither as a *peacemaking* nor as a *peacekeeping* force, but for the protection of convoys and released detainees. Thus, it was inevitably a spectator to the fighting and, at least in theory, could only take military action in defence of its own soldiers' lives, UN property or the people it was protecting.

From this Stewart developed a 'concept of operations' which was much more positive in nature. Since he considered that it was insufficient simply to place armed escorts at the front and rear of each convoy, he believed he should try to ensure through a mixture of diplomacy, cajolery and persuasion that areas where aid was being delivered were as peaceful as possible. Such active measures would, in his view, not only ensure that aid was delivered safely, but might also, in the longer term, result in a reduced need for aid.

As a consequence Stewart found himself undertaking many other tasks, principally almost incessant negotiations and conferences. These frequently concerned convoys, but at other times they involved issues such as refugees. The battalion also took military and civilian casualties to hospital and recovered dead bodies, while the engineers were involved in clearing and improving roads and bridges, to the benefit of both UNHCR convoys and the civilian community.

The battalion also undertook a task, intangible in nature but best described as 'presence', which exploited the fact that when UN troops were actually within sight extreme action by one faction or another was unlikely – in other words, as Stewart put it, 'If there's trouble, get into the middle of it and calm things down by being there.'

In all this, it was essential not only that the force saw itself as neutral, but also that it was seen by all sides to be so. Stewart's visits had to be divided equally between all parties and he even went so far as to ensure that invitations to social occasions were equally balanced and that sporting fixtures were played against each side in turn (and that, on every occasion, the British just lost).

On the ground, the most important of the battalion's tasks was the protection of aid convoys. These started as soon as the Cheshires arrived, with the first convoy to Tuzla running on 19 November. The network of routes expanded rapidly and by December the battalion was protecting six to eight convoys a day, including a weekly convoy from Belgrade to Tuzla. However, the task of escort involved much more than simply a physical presence. Stewart needed to create the conditions necessary for free movement, one particular hazard being the large number of checkpoints maintained by all sides – permission was needed to pass through each, and petty-minded guards could impose serious delays simply in order to humiliate the United Nations.

Off the main routes, the country roads and bridges, not particularly good even in Tito's day, had seen little maintenance since. The arrival of the UNHCR, with its heavy trucks, many of them articulated, made considerable demands on the routes and it was therefore necessary for the battalion group's engineers to devote considerable efforts to maintaining and improving them.

The Cheshires came under fire regularly from small arms, 12.7mm machine-guns, 40mm Bofors, mortars, rocket-launchers and artillery. Mines were also an ever-present danger. A key element in overcoming this was the Warrior, which not only provided armoured protection for its occupants but was also large and impressive. The Warrior caused some problems because of its weight (for example, when crossing bridges) and width (when trying to pass down narrow streets) but its advantages far outweighed its drawbacks. In addition to the protection it afforded, it also mounted significant weapons of its own: one 30mm Rarden cannon and a Hughes 7.62mm chain-gun, both of which were fitted with day and night sights.

Despite such protection, the Cheshires lost one man, the driver of a Warrior, and a number were wounded, the majority shot in the head by deliberately aimed sniper fire. Medical backup, however, was excellent, with a field surgical team at Vitez and smooth evacuation procedures. For example, one man wounded in the early afternoon in Gornji Vakuf was extracted by helicopter to Split and flown to Britain, reaching hospital in London by 2200 that evening. Other hostile action against UN forces in the BHC area included hijackings and hostage-taking, although the Cheshires were not involved.

Unlike in normal military operations, the Cheshires had to contend with three combat forces and a civilian population, and to deal with many other agencies in the area. The most important of these was the International Committee of the Red Cross, but others of immediate relevance to the unit's role included UN military observers and the European Community Monitoring Mission, both of which were unarmed and operated in small teams.

In addition, there were many charities. Some were well-organized general bodies like the French Pharmacies sans Frontières, some groups had agendas of their own such as the US-based International Rescue Committee; and there were smaller charities, often established to deal with a specific element of the conflict, for example, the Bosnia Winter Appeal. While operating independently, many of these looked to Unprofor for help when they got into difficulties. There were also many government organizations operating in the area, such as the British Overseas Development Agency.

Other combatants occasionally came to the Cheshires' notice. These included mercenaries, who were present in small numbers. Generally given little time by anybody, including their employers, they were used for jobs of limited responsibility and they were as likely to be killed by their own side as by their enemies. It was also rumoured that there were volunteers from Islamic countries, although their presence was not confirmed until later. Finally, there were local gangsters who took advantage of the conflict for their own criminal ends.

The press was an omnipresent factor. The British media naturally had an interest in the activities of the Cheshires. So too did the US press, for in the absence of any direct US involvement the British had the advantage of speaking English. Stewart recognized from the start not only that the press had a job to do but also that, if handled properly, it could be made to work to his advantage. As a result, he applied a number of principles, among which were being helpful at all times, briefing the press wherever possible, and including them in his travel plans. He also gave personal interviews when requested to do so both to the international press and to the local press, including TV stations. In this he was assisted by a British PInfo team based at Vitez. As a result, the Cheshires received a favourable press throughout their tour, as did Stewart himself. He came across well on TV and was generally perceived to be an honest man, by no means 'stuffy', and one who was quite prepared to speak his mind. One unfortunate consequence was that there were elements within the British Ministry of Defence who considered him to be developing something of a 'personality cult', although he was never formally warned to curb his activities.

Unlike most other examples in this book, Stewart did not command in any one battle. Instead, the six-month tour involved a continuous stream of different challenges, each requiring its own unique solution. One example (among many) was the cessation of the Serb bombardment of Turbe and Travnik over the Christmas period. This originated with Stewart's desire to seek Serb agreement to a crossing point for convoys at Turbe, which would considerably shorten their route. The process was set in train when he discovered that a Travnik-based Muslim officer, Captain Beba, was the Muslim/Croat representative on an exchange commission which met at Turbe to arrange the transfer of prisoners and bodies with the Serbs. Stewart asked Beba to arrange for him to attend a meeting, to which the Serbs agreed, their only stipulations being that no Unprofor armoured vehicles were to cross the lines and that if shelling was taking place the meeting would be aborted.

The first meeting was held on 17 December 1992 when, as agreed, a local ceasefire started at 0800. Then, at 1000, in a show of military force which Stewart always found helpful in establishing his credentials, four Warriors entered Turbe to ascertain that all was quiet. When this was confirmed, the meeting party drove in. It consisted of Beba in a civilian car flying a large white banner bearing a red cross, and three British Land Rovers carrying Stewart and his party. This group crossed the Muslim/Croat front line at precisely 1100, the distance between the two lines being about 1,000 metres and including several rows of mines lying on the road, which were moved by pushing them to one side. The progress of the convoy was watched at close range by heavily armed soldiers of both sides but the convoy passed through the Serb front line without incident and arrived unharmed at the café where the meeting was to be held.

The Serb delegation consisted of six captains, and Stewart started the meeting by outlining his mission, stressing that the UNHCR wished to bring aid to Serbian as well as to Muslim and Croat areas. He then requested a meeting with Serb commanders. This first meeting served to break the ice and ended with an agreement to meet again in seven days' time. At that second meeting the same Serbs were present but told Stewart that his request to meet senior Serb commanders would have to be passed through UN channels to the Serbian army's headquarters at Pale. (This was done, but the UN never received a reply.) A third meeting took place at Turbe on Christmas Eve and followed the previous pattern of a local ceasefire, although this time the first item of business was for the British to help recover the corpse of a Serb soldier lying in no man's land and return it to the Serb lines. At the meeting the Serb delegation was led by a major, the deputy commander, 19 (Serb) Brigade. Stewart, never one to let an opportunity pass by, asked for a local ceasefire from the end of the meeting until the Serb Christmas on 8 January. The Serb major replied that his army never fired on civilians (a palpable untruth), but that he would agree to such a ceasefire if all other sides did so. Stewart quickly obtained Muslim/Croat agreement and managed to get significant parts of the meeting filmed by the BBC, making it difficult for the Serbs to retract their promise had they wished to do so. As a result, several thousand civilians had a period of over two weeks free from artillery and mortar fire – a victory of sorts in a confused and bitter war.[3]

Many peacekeeping operations have taken place in the past fifty years and each has had its own unique characteristics, making it very difficult to compare one with another. Thus, Stewart's problems and actions in Bosnia bear no comparison with United Nations operations in Rwanda, East Timor or Sierra Leone, nor was there any adequate precedent upon which he could base his operation. His policy of direct involvement gave him a high personal profile, while his deliberate decision to engage in close relations with the media meant regular exposure, especially on television.

Under modern conditions, what is normally considered to be 'peacetime' can suddenly be disrupted by a small group which strikes at a time and place of their own choosing. Although some general precautions can be taken, such actions are by their very nature unpredictable. Thus, for example, the sudden seizure of the United States Embassy in Tehran by a group of Iranian students on 4 November 1979, the raid on the Iranian Embassy in London in April 1980 by anti-Khomeini terrorists, and the hijacking of the cruise ship *Achille Lauro* by Palestinians in October 1985 were carried out by totally unrelated groups who employed tactics which, in each case, were entirely novel. And even when familiar tactics such as hijacking are used, the perpetrators can create surprise by their choice of time and place, as, for example, in the case of the hijacking of the Air France flight AF 139 on 27 June 1975. On that occasion, though the flight started from the Israeli airport at Tel Aviv, a mixed group of Palestinian and Baader-Meinhof terrorists were able to redirect the flight to Entebbe in Uganda. Similarly, the seizure of the *Mayaguez* could not have been foreseen.

A very small number of the post-Cold War campaigns have been conducted along traditional military lines, the major example being the Gulf War, but the remainder have been relatively small-scale operations, frequently under United Nations' auspices, including those in Bosnia, Croatia, East Timor, Kosovo, Rwanda, Somalia and, most recently, Sierra Leone. Not only have these been of a new nature but each has involved a unique combination of political, military, ethnic and religious factors that make it difficult to relate them to each other and to deduce common patterns.

Three issues do seem clear, however. The first is that the need for military involvement in these peacekeeping and humanitarian missions arises because other means of solving the problem have failed. This means that, by definition, the problem is not subject to an easy solution and that virtually every other solution has already been tried. In such situations external military force is normally intended to provide little more than a short-term cooling-off period by keeping warring parties apart or by preventing atrocities of too horrific a nature, while politicians seek a more enduring solution. Unfortunately, such solutions are not always forthcoming, with the result that the peacekeeping force may remain for many years, one example being the UN force in Cyprus which has been keeping the Greek and Turkish Cypriots apart since 1964. That is why Stewart and the Cheshires, like so many other UN forces, could never have been expected to find a solution to the problems in their area of Bosnia.

Second, it is clear that in such circumstances the commanding officer is the most important link in the chain of command. He obviously cannot be present at every incident, and thus captains, lieutenants, sergeants and corporals frequently have to take the decisions on the ground, usually quickly and often with the press recording their every move. Similarly, senior officers, both national and international, also have a role to play, though it is essentially remote and advisory. It is nevertheless, the commanding officer who sets the agenda and defines the framework, both operational and ethical, in which his officers and soldiers operate. Members of the Cheshires were direct witnesses of atrocities beyond their wildest imagination but were seldom in a position either to prevent them or to bring those responsible to justice. It was their commanding officer who sustained them through such ordeals and who ensured that, if they did react, it was within the close limits set by the UN mandate and national rules of engagement.

Finally, such cases involve moral dilemmas. In Bosnia, the movement of refugees who had been displaced by Serbian 'ethnic cleansing' was at one level a form of complicity. At another, more pragmatic, level the perpetrators were determined on their action. By transporting the refugees Unprofor was at least ensuring a safe passage for them. Other chapters in this book deal with some of the moral problems which can face commanding officers in general war, but such difficulties tend to arise only periodically and they are fairly easy to recognize. In Bosnia moral questions arose on an almost daily basis. It is therefore perhaps worth quoting the criterion which Stewart set himself: 'I would always ask myself whether a course of action was essentially good or bad. If I felt it was morally defensible, then I would consider doing it . . . I looked at problems in simple terms of whether an action was essentially right or wrong.'[4]

12

The Loss of HMS *Fittleton*, 20 September 1976

The great majority of armed forces depend on reserve forces to complete their order of battle. There are numerous such organizations in the United States, including the Navy, Army, Air Force and Marine Corps Reserves, and the Army and Air National Guard, while in the United Kingdom the major volunteer reserve forces are the Royal Naval Reserve, the Territorial Army and the Royal Auxiliary Air Force. All such forces depend upon their male and female volunteers to agree to be called up in the event of war or disaster and to undergo regular training in peace to prepare them for such contingencies. The actual amount of training varies between nations and forces but typically includes one evening per week, one weekend per month and, in most cases, at least seven days' continuous training every year. During such training the reservists practise for their war roles, usually on exercises. They dress and train as regulars. They also bear the same responsibilities, as Lieutenant-Commander Peter Paget of the Royal Naval Reserve (RNR) discovered when he was court-martialled following the sinking of HMS *Fittleton*, the minesweeper he commanded, on 20 September 1976. Twelve lives were lost in what was, the Falklands War apart, the Royal Navy's most serious surface disaster during the Cold War.[1]

The chain of events is here described in some detail in order

both to explain what happened and to understand the responsibilities placed upon the commanding officer.

In the mid-1970s, the RNR had a strength of 5,600, and those
who served within it came from a wide spectrum of society.
Some were former regulars who had served in the Royal Navy
and others were professional civilian mariners, but most came
from jobs and backgrounds that had little to do with the sea.
Nevertheless, the standards set were the same as those for regulars. The reservists had to pass the same examinations and
tests, and to undergo a specified number of days training,
which included an annual period at sea, normally in the
summer months.

The most important of several missions assigned to the RNR
was minesweeping, a highly specialized activity in which the
reservists' expertise and skills were acknowledged as being at
least as good as, if not in some cases better than, those of
regular naval crews. Their main equipment for this work was
the Ton-class minesweeper which, by 1976, had been in service
with the RNR for twenty years.[2] RNR ships' captains and
crews were used to close-quarter work, since minesweeping
duties routinely included passing sweep wires from one ship
to another, while equipment and stores were also regularly
transferred using jackstays slung between kingposts, which
were mounted amidships just forward of the funnel. The minesweepers also regularly carried out heaving-line transfers, in
which two ships approached sufficiently close for a seaman on
the forecastle of one to throw a line to land on the forecastle of
the other.

From 1958 to 1974 the RNR was organized into eleven
administrative divisions around Britain, with each division
operating one Ton-class minesweeper. These were used for
training in peace, but were then concentrated in peacetime
exercises (as they would be in time of war) to form the Tenth
Mine Countermeasures Squadron (10th MCMS). In 1974,
however, the RNR suffered particularly badly in a defence
economy drive. Though the number of RNR divisions was
maintained at eleven, the number of minesweepers was

reduced to eight. This meant that three divisions lost their minesweepers and that a process known as 'hull-sharing' had to be introduced. Specifically, in south-east England, London Division lost its ship and then had to obtain 'time slots' with the two adjacent divisions for the use of either HMS *Crofton* of Solent Division, which was based at Southampton, or HMS *Fittleton* of Sussex Division, based at Shoreham in Sussex.

Despite the reduction in the number of ships, the number of statutory sea-training days remained unchanged, which resulted in some serious programming problems for the RNR planners, especially for the London Division, and meant that individual officers and men received less sea-time each year.

The larger of the two ships involved in the incident about to be described was HMS *Mermaid*, which had a curious history. She had been ordered in 1957 as a combined state yacht/training ship by President Nkrumah of Ghana, and to save time and expense much of the design was based on that of existing British frigates. However, the superstructure was quite different, since the Ghanaian Navy required large staterooms, all on the upper deck, which pushed the bridge forward, resulting in a smaller than usual forecastle, most of which was taken up by a twin 4-inch gun turret.

When Nkrumah was ousted by a military coup in February 1966 the contract was cancelled and, after languishing in the Clyde for several years, the vessel was purchased by the Royal Navy and commissioned in 1973 as HMS *Mermaid*, for employment as a training ship. She was generally considered to have served her purpose well but, though most of her equipment matched standard naval items, she was always a 'one-off' with many peculiarities. After only a few years' service she was sold to Malaysia.

HMS *Fittleton* was one of a larger number of Ton-class minesweepers which were built as part of a Nato programme in the 1950s and 1960s and which were intended to counter the very serious threat posed to Nato shipping by the latest development of the Soviet magnetic influence mine, which could not

TABLE 14: *HMS* Mermaid *and HMS* Fittleton

Ship	HMS *Mermaid*	HMS *Fittleton*
Role	Training frigate	Minesweeper
Commissioned	16 May 1973	28 January 1955
Displacement (deep load)	2,520 tons	440 tons
Dimensions		
Length	339 feet	153 feet
Beam	40 feet	29 feet
Maximum speed	24 knots	17 knots
Complement	210	29
Weapons	2 x 4-inch guns 2 x Bofors 40mm 1 x mortar	1 x Bofors 40mm 1 x twin Oerlikon 20mm

be overcome by the degaussing methods then in use. They were extremely sturdy ships with a wooden hull made of a double layer of mahogany laid on an aluminium frame, and the Royal Navy took delivery of 115, twelve of which were originally allocated to the RNR.[3] They were highly manoeuvrable and seaworthy ships, their only problem of significance to this story being that their Napier Deltic diesel engines consumed large quantities of lubricating oil. This could result in an accumulation of partly burnt oil in the exhaust system and funnel which, under certain circumstances, could be ignited, resulting in a fire. For technical reasons, this problem was at its worst at a speed of 11 knots, so an Admiralty order was issued in the late 1950s instructing ships to avoid this speed if at all possible.

The official complement for a Ton-class minesweeper was twenty-nine officers and ratings, ideally all from the same RNR division in which they would have trained together both ashore and afloat. However, various factors dictated that this was not the case aboard *Fittleton* on 26 September 1976. First, the RNR reorganization and 'hull-sharing' had caused the problems already described. Second, London Division suffered shortages in specific trades, particularly radio operators and cooks. Third, there was a degree of overmanning in order to give as many individuals as possible the opportunity to fit in the necessary sea-time, which was obligatory to qualify for the annual bounty. Finally, *Fittleton* was the senior ship and carried the Commander 10th MCMS and his yeoman of signals. As a result, at the time of the accident *Fittleton* had forty-four people aboard, of whom six were regular sailors on loan to fill vacancies for which no RNR men were available. It might be thought that regular sailors would have been more experienced than their RNR shipmates, but four of the six were very young men who had just completed their training and who were awaiting posting to their first ship. They thus had considerably less sea-time than the majority of the reservists.

During the events about to be described, *Fittleton* was commanded by Lieutenant-Commander Peter Paget, RNR. Aged 44, he was an experienced Merchant Navy officer who had served for many years in the large and prestigious P&O shipping company and who held a full Merchant Navy Master's Certificate of Competence. He had served in the London Division for sixteen years, almost all of it aboard Ton-class minesweepers. He thus had more experience of minesweeping and more sea-time than many regular officers of his age and seniority, was thoroughly versed in the ways of the Ton-class, and was respected by reservist colleagues for his well-proven seamanship and leadership qualities.

Nato's Exercise Teamwork started on 12 September 1976 with 300 warships taking part, of which the 10th MCMS contributed seven Ton-class minesweepers, four (*Crichton*, *Crofton*, *Fittleton* and *Hodgeston*) operating in the English Channel, and three (*Kedleston*, *Repton* and *Wiston*) in the Moray Firth. The plan was that these ships would sail from their home ports on Friday, 10 September, take part in the exercise from 11 to 17 September and then disengage in order to steam towards a

12. The *Fittleton* Disaster, 20 September 1976

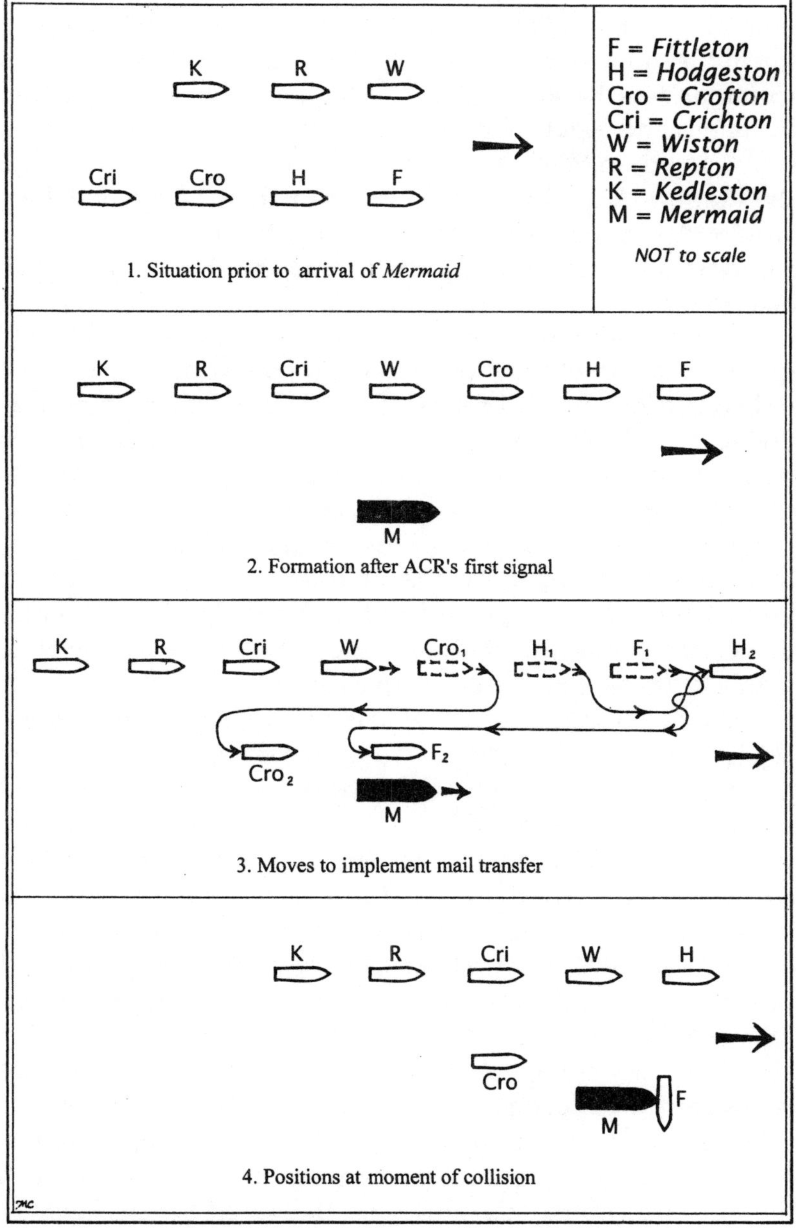

rendezvous in the North Sea on Monday, 20 September. The senior officer was Commander Ian Berry, RNR, who was aboard, but not in command of, *Fittleton*. Once at the rendezvous the minesweepers would be joined by the Admiral Commanding Reserves (ACR), a regular rear admiral, flying his flag in the frigate HMS *Mermaid*, which would come south from Rosyth in Scotland. Once all eight ships had joined up, the admiral would assume command from Berry, and they would then proceed to the German port of Hamburg for a 'courtesy visit', returning to their home ports on Sunday, 26 September.

Following successful completion of the exercise, the two groups made their way towards the rendezvous, but at about 0945 on 20 September Berry received a signal telling him that, owing to engine trouble, *Mermaid* was behind schedule. He was to steer a course for the mouth of the River Elbe at 10 knots and *Mermaid* would overtake him in the early afternoon. The minesweepers duly met off the east coat at 1428. Berry formed them into two columns, one of four ships (the Plymouth group), the other of three ships (the Tyne group), and continued on the course directed, carrying out 'officer-of-the-watch' manoeuvres as they went (see diagram 12.1).

The minesweepers were still in the same two columns when at 1451 *Kedleston*, the rearmost ship, signalled Berry in *Fittleton* that *Mermaid* had been sighted astern and was approaching at a speed of between 15 and 20 knots. As the minesweepers were on a slightly different track to *Mermaid*, Berry ordered a 45° change of course for the whole formation, but the change had no sooner been made than he received two voice messages from the admiral. The first was a communications check. In a second, much longer message the minesweepers were warned that the admiral would shortly be ordering them to form a line ahead on a course of 100° and at half standard distances (that is, 760 feet). Two further elements of this warning order caused a little surprise among some of the minesweeper captains. The order laid down for the ships was different from that in which they were currently steaming, and the speed ordered was 11 knots, a speed on which, as has been noted, there was a long-standing prohibition.

This signal was passed by voice radio, using codes from the Naval Signal Code Book, and it took several minutes for the signallers aboard each minesweeper to decode the message and then inform their captains of what was required. Berry considered that, as the senior officer currently in command, he was responsible for conducting the group until the ships were disposed in the formation required by the admiral, at which point he would hand over command. He therefore ordered a reversal of course and started to lead the two columns back to meet the advancing *Mermaid*, planning to merge the two columns into one in the order stipulated by the admiral as they approached the flagship. He would then order a second course reversal to bring the column on to the same course as the flagship, at which point command would pass to the admiral. Before this could be achieved, however, the admiral sent a further signal ordering the immediate execution of the previous signal, which not only scotched Berry's plan but also resulted in a mêlée as the ships moved across and around each other, and then reversed course in order to comply with the admiral's instructions.

The minesweepers managed to sort themselves out without major incident, and were soon in line ahead on a course of 100°, with *Mermaid* steaming on a parallel course 14 cables (2,800 yards) to starboard of the central ship, *Wiston*, as shown in diagram 12.2. At this point the admiral issued a further signal, also a warning order, and also lengthy and using the Naval Signal Code Book. This signal instructed all seven ships to close on *Mermaid* one after another for a heaving-line transfer, *Fittleton* leading, with the next in line moving to a 'waiting position' 6 cables astern of the flagship.

Such an evolution had been forecast in a signal from the admiral earlier that day and there was nothing immediately unusual about it. It was generally assumed aboard the minesweepers that the admiral had two purposes: first to pass them the written orders concerning the Hamburg visit, which they had not yet received; and second, to afford him an opportunity to observe each ship at close quarters and while conducting a manoeuvre that would test both the captains' and the crews' seamanship. The minesweeper captains, were nevertheless surprised that the speed ordered for this manoeuvre was also set at 11 knots.

The meteorological and sea conditions for the planned manoeuvre were ideal. There was plenty of sea-room, the depth of water was about 120 feet and there were no known navigational hazards. The weather was good, with visibility of 5 to 7 miles, no bad weather was forecast, and sunset was not due until 2000. The wind was Force 3–4, with a slight sea and waves about 3 feet high, but these conditions would not affect the transfer.[4] Finally, there were no other ships in the vicinity that might interfere with the planned manoeuvres, although there was a West German frigate, FGS *Bayern*, some distance away engaged in other business.

Again, it took several minutes for the signallers to decode and acknowledge the signal, but as soon as it was realized what was intended each ship piped 'close-up special sea-duty men' in order to carry out the preparatory actions dictated by standing orders. This involved most off-duty men coming on watch and moving to their places of duty, both above and below decks, which was necessary for both safety reasons and to deploy the men required to conduct the planned transfer to their correct stations. One of the specific requirements of this order was that the first lieutenant, the chief boatswain's mate and a group of seamen fell in on the forecastle, where they would be responsible for conducting the actual transfer. In the engine room, the engineer officer and two of his artificers were required to go to the engine control room, while another two members of the engine-room staff had to proceed to the bridge where they would be responsible for synchronizing the revolution counters on the bridge with those in the engine-room (this had to be done twice, once for each engine). At that point control of the engines would be formally transferred to the bridge. Other men also took up their assigned places of duty,

This activity was accompanied by a raising of the Damage Control State throughout the ship, which included closing the watertight doors, a precaution against the consequences of collision or grounding.

Clearly, each ship required a few minutes to complete these actions, the most critical being the first ship, *Fittleton*, but these activities had not been completed aboard that ship when the admiral sent a further signal, demanding 'What are you waiting for?' At this point, it was clear to Berry that the admiral had assumed command, which he signified to Paget by moving off the bridge and on to the flag deck. From that moment he and his yeoman were, in effect, passengers. Unfortunately, this 'hastener' signal had been incorrectly coded aboard the flagship, with the result that it was not immediately clear whom it was intended for. Paget assumed that it was meant for him and so, despite not having completed the transfer of engine control to the bridge, he started his turn to starboard, only to discover to his astonishment that both *Crofton* and *Hodgeston* were also wheeling. Thus, Paget had to take immediate action to avoid one of the minesweepers which was crossing his path. He ordered the starboard engine stopped to reduce speed until the other ship was clear and then carried on towards the flagship, as shown in the diagram on p. 221, completing transfer of control of the engines in the process.

Astern of *Fittleton*, the captain of *Hodgeston* realized that he had misunderstood the admiral's signal and turned back on to the original course. The captain of *Crofton*, however, having avoided first *Fittleton* and then *Hodgeston*, decided very quickly that, rather than try to resume his place in the line, he would cause the least confusion and danger by carrying on after *Fittleton* towards *Mermaid* and then moving into the 'waiting position' where he would be the second to carry out the transfer. This piece of quick thinking was possibly not a 'text book solution' but it was the best course in the circumstances and, as it would turn out, very fortunate for those aboard *Fittleton*.

Having successfully avoided the other two ships, Paget continued to head towards a point on *Mermaid*'s beam. Once in the correct position he turned to port to bring his vessel on to a course parallel to that of *Mermaid*, which was then on his starboard beam and 100 yards distant. Paget could see *Mermaid*'s sailors waiting to receive the line at the transfer position, marked by a green flag and just forward of her port bridge wing (it was, in fact 68 feet from the bow),[5] while aboard *Fittleton* the heaving-line party was standing by on the foredeck immediately forward of the minesweeper's bridge and against the starboard rails. Seeing that *Fittleton* was a little behind the required position, Paget increased his speed fractionally to 12 knots to pull his ship slowly forwards and then started to edge closer towards *Mermaid*.

The standard naval heaving line is 17 fathoms (102 feet) long, and according to the *Manual of Seamanship* most seamen overestimate their ability to throw it – 12 fathoms (about 70 feet) is considered a good cast.[6] Paget, however, brought his ship in closer, presumably because he suspected his part-time sailors were out of practice, but also because *Mermaid*'s forecastle was considerably higher than his own. Even so, when the first cast was made at a distance of 40 feet, the end of the line hit *Mermaid*'s side and fell into the water. The thrower quickly heaved his line in, in order to recoil it in preparation for a second attempt.

Paget was aware of this failure when suddenly, and without any adjustments to the helm or engine controls, *Fittleton* was sucked in sideways towards *Mermaid*. She hit broadside on and then bounced off. As she did so she also veered to port, with the bow moving outwards and the stern moving inwards towards *Mermaid*. Berry, who was standing on the flagdeck, quietly told Paget to 'watch your stern'. Paget ordered 'helm amidships', followed by 'starboard 20 degrees', with the object of pulling the stern off. This succeeded in straightening the ship, but then, without any alteration to her engine speed, *Fittleton* began to move forward relative to *Mermaid*, giving Paget the distinct impression that she was surfing. His order for a reduction in speed to 10 knots had no effect as *Fittleton* continued to surge ahead.

At this point Berry suggested to Paget that he might try to break free by increasing speed and steering to port, that is, away from *Mermaid*, but by then *Fittleton* had drawn abreast of *Mermaid*'s bow. At that point she suddenly, and apparently of her own volition, swung around 90° to starboard, directly into *Mermaid*'s path. The frigate's captain immediately ordered 'full astern' but it was too late. The larger ship rammed *Fittleton* amidships, her bow pushing the smaller ship over on her port side.

These events were so sudden and unexpected that no one on *Fittleton*'s bridge had time to warn those below of what was about to happen. Within seconds the ship had reached an angle of about 70 degrees, at which point those on the upper deck were thrown into the water. The ship continued to roll, pushed ever further by *Mermaid*'s momentum and almost hitting those already in the water, until she came to rest upside down but still afloat. The position of all ships was now as shown in diagram 12.4.

Those below at the time were placed in a terrible predicament. They had been given no warning of impending disaster and found themselves, suddenly and without explanation, with the deck where the deck-head should have been, equipment and debris falling all around them, and their compartments flooding very rapidly until only small air pockets were left. Some managed to work out an escape route, which involved swimming underwater along passageways, occasionally finding an air pocket for a quick gasp of air, before continuing until they reached what would, under normal circumstances, have been the upper deck, when they headed for the surface.

At the moment of impact *Mermaid*'s captain brought his ship to a very rapid standstill and then stopped the propellers in order to avoid injuring any survivors in the water. Fortunately, *Crofton*, which had been in the 'waiting position', now found herself very close to the upturned hull which had passed down *Mermaid*'s starboard side. *Crofton*'s captain stopped engines to allow his ship to drift down among the survivors. Some were pulled aboard on ropes, while others were rescued by two small boats which had been lowered into the water. The West German frigate *Bayern*, having seen that something was amiss, came up very quickly and sent her medical officer to *Crofton*, where he helped to treat the shocked survivors. Altogether thirty-two survivors were recovered and by about 1600 all had been transferred to *Mermaid* which, being much larger, was better able to handle such numbers.

Most of the men who had been on the upper deck, including Paget and Berry, were rescued, and a number of those trapped below decks managed to escape. Two bodies were also recovered. That left ten men unaccounted for. However, it seemed possible that some survivors might still be trapped in the air pockets that were keeping the hull afloat. An officer was put on to the hull and when he tapped he heard similar tapping noises from inside, but whether these came from survivors or from equipment moving about could not be established.

The anxious men on *Mermaid*'s bridge now considered whether to attempt a direct rescue or to keep the hull afloat pending the arrival of more capable, specialized rescue ships. Direct rescue could be effected in two ways. Divers could enter the hull from underneath and then climb through the ship to find any survivors and help them escape. The problem with this solution was that if *Fittleton* sank while they were inside, the divers would go down with her. The second possibility was to cut a hole in *Fittleton*'s exposed bottom, which would give access but might simply result in the escape of air that was keeping the hull afloat. It was also unclear how the hole could be made since *Fittleton*'s hull was constructed of a double layer of mahogany which was designed to resist the explosion of a mine and for the puncture of which no suitable tools were available.

Yet another possibility was to keep the hull afloat. Again there were two ways in which this might be accomplished. A kite wire (used in sweeping operations) could be run between two other minesweepers which would then allow themselves to run down on *Fittleton* until the wire snagged, whereupon they would turn outwards and maintain the wire at sufficient tension to keep the hulk afloat. This was tried between 1700 and 1830 but proved unsuccessful. It was also suggested that a cable be secured aboard *Mermaid* and that a motor-boat should then pass the free end under the bulk and return it to the frigate where it would be secured, thus using the much larger ship to keep the minesweeper afloat. Again there were doubts about the strength of the cable and of the holding points aboard *Mermaid*. These discussions on *Mermaid*'s bridge were protracted, having started at least as early as 1600 and lasting until 1830. They became academic when *Fittleton* suddenly sank stern first and came to rest on the bottom.

At 2100 another British frigate, HMS *Achilles*, arrived, to which the admiral transferred to continue his responsibility for the recovery operation. *Mermaid* was released to take the survivors back to England, while the Ton-class ships eventually proceeded to their home ports, their eagerly awaited 'run ashore' in Hamburg having, naturally, been cancelled. At dawn the following morning British and Dutch naval divers took it in turns to go down to the hull, but knocking went unanswered and any hope of finding further survivors was abandoned. Recovery vessels raised the hulk some days later and took it to a Dutch port where it was dried out and three bodies removed. The hulk was then returned to England where a few months later it was scrapped.

Seven bodies were never recovered, but a civil inquest was held at Chatham on 11 January 1977 into the deaths of the five men whose bodies had been brought back to England for burial. The inquest recorded a verdict of accidental death. Meanwhile, a Board of Inquiry, the first step in the Navy's legal proceedings, was held 'in camera'. Its proceedings have never been published, but it must be presumed that it found that a *prima facie* case against Paget had been established.

Initially there were some doubts as to whether the naval custom of court-martialling a commanding officer who had lost his ship would extend to a reserve officer, but these were dispelled when a court martial convened at the naval base in Portsmouth on 20 February 1977. The court was presided over by a captain (the commanding officer of the Royal Navy's largest warship, the aircraft-carrier HMS *Ark Royal*), and consisted of four members (none of whom were reservists). A naval captain acted for the prosecution and Paget was represented by a reserve officer who was also a practising barrister.

Paget faced four charges. The first was 'negligence in altering course towards *Mermaid* when it was not justified'. The others concerned 'failure' in three respects: first, 'to take early and positive action to correct a dangerous alteration of course towards *Mermaid*'; second, 'to ensure that his orders to the wheel were obeyed, which caused *Fittleton* to be lost'; and third, 'to reduce speed and alter course to extricate his ship from a dangerous position while she was close alongside the frigate *Mermaid* during a mail-transfer operation'.

The proceedings opened with the prosecution submitting a 'circumstantial letter' which set out the facts of the case and then calling five witnesses, of whom two are germane here. The main prosecution witness was the admiral. He stated that when there was no apparent reaction from *Fittleton* to his signal to start the heaving-line transfer, he had sent the signal: 'What are you waiting for?' He accepted, however, that it had been incorrectly 'coded-up', which resulted in the three leading minesweepers all pulling out of the line virtually simultaneously. When the defending officer asked the admiral if he would agree that there had been a muddle over the signal, the admiral replied, 'If the captain of *Fittleton* was unhappy about the situation, he could have hauled down his signal indicating that he was not ready to go up [that is, to start the evolution]', adding that '*Fittleton*'s captain need not have made the approach until she was ready. There was no need for her to have moved into that position until she was ready to do so.' He ended by saying, 'If any of the RNR commanders were unhappy about an exercise because of inexperience, I would have expected them to say so'.

A serving frigate captain was called as an 'expert' witness on matters of seamanship, and part of his evidence is worth quoting in full:

> We have heard that *Fittleton* made a marked alteration of course away from *Mermaid* immediately after the first collision. The result of that alteration was to swing his stern sharply towards *Mermaid*'s side. In order to prevent a second collision the CO of *Fittleton* checked the swing and paralleled the *Fittleton* up again with *Mermaid*. This in itself would not have been unseamanlike had not *Fittleton* continued to accelerate up *Mermaid*'s side. The most vital need of this moment was that *Fittleton*'s speed should have been reduced to that of *Mermaid* or even below. During those vital seconds *Fittleton* had continued to move ahead into that very dangerous position under *Mermaid*'s bows. He [Paget] should have made a series of small alterations away having first of all matched his speed with that of *Mermaid*, which we know was proceeding at 11 knots.

When asked by the defending officer whether Paget had been negligent, this witness replied: 'In my opinion he had time to have acted earlier in preventing that first collision.'

The main facts – that is, the events and the sequence in which they happened – were not disputed, but at the end of the case for the prosecution the defending officer submitted that there was no case to answer. His grounds were that the prosecution case was so weak that no reasonable court could convict the accused, and it was quite clear, in his view, that *Fittleton* had been 'sucked towards the bigger ship by underwater currents' and not steered in. After consideration, however, the court decided that there was a case to be answered and requested the defence to present its witnesses.

When Paget presented his case he picked up the story at the point when *Fittleton* first hit *Mermaid*:

I considered I had made a mess of it and I was determined to have a second shot . . . we closed gently. My intention was we should move in and settle. However, that did not happen and as we approached to 40 feet of *Mermaid*, the heaving line which was thrown bounced off *Mermaid*'s guard rail. Seeing we were not parallelling up I immediately ordered 20 degrees to port. The ship moved bodily sideways and when we were 20 feet off she began to move ahead. I instructed the engine room to reduce speed by one knot. The ship then began to straighten and continued to move ahead until we were virtually parallel but some eight to 10 feet away. We struck *Mermaid* not particularly violently and bounced and tilted slightly to port. My attention was drawn to the fact that our stern was closing on *Mermaid* extremely rapidly. There was no doubt in my mind we would inflict severe damage on *Mermaid* and ourselves. I ordered midships and starboard 20 degrees with the object of pulling the stern off. It did have the effect of stopping the swing and we found ourselves parallel with *Mermaid*. Having resolved that comparatively minor crisis I realised that *Fittleton* was continuing to move forward. I ordered a further reduction in speed to 10 knots. I was not entirely conscious of what was going on for perhaps three to four seconds, and as the ship was moving violently to starboard, I could only describe it as a blind spot due to a disorientation arising out of the speed of the slew to starboard. I saw *Mermaid* catching the top of our gunwale, and with ourselves moving forward I thought we were clear . . . my own recollection is that she did not start to heel until she hung over.

Paget added that he was conscious of the lack of directional stability and that he had the sensation that the ship was surfing.

Having heard all the witnesses and the closing submissions, the court closed to deliberate. It reopened several hours later to announce its verdicts, which were that on three charges

Paget was not guilty but that he was guilty of the fourth charge of failing 'to reduce speed and alter course to extricate his ship from a dangerous position while she was close alongside the frigate *Mermaid* during a mail-transfer operation'. After hearing evidence as to character the court sentenced Paget to a reprimand, the lightest punishment available to it.

The entire proceedings were then submitted to the Admiralty Board for confirmation, but on 11 May 1977 it was announced that:

> The Admiralty Board of the Defence Council has reviewed the proceedings [of the Paget court martial] and has concluded that while Lt-Cdr Paget hazarded *Fittleton* due to errors of judgement based on lack of experience, these did not amount to negligence under section 19 of the Naval Discipline Act. In these circumstances the Admiralty Board has decided that the conviction is unsafe and unsatisfactory and has accordingly quashed the conviction and annulled the sentence.[7]

The key question in this case was why *Fittleton* suddenly swerved to starboard and under the bow of the much larger frigate. It had long been known that the forward motion of a ship causes hydrodynamic disturbance, which manifests itself visually on the surface as 'bow waves' and 'wake', and that the areas around the bow and stern are areas of positive pressure, with a corresponding area of (negative) suction between them. There had also been a number of recorded incidents in which ships had apparently been 'sucked' together when they had been steaming on the same course and in close proximity. This phenomena was termed 'interaction' and it was also known that its effects could be aggravated if there was a significant difference in size between the two ships.

Two such incidents had occurred in the Second World War. The first involved the cruiser HMS *Euryalus* and the destroyer HMS *Worcester*. They were steaming on parallel courses 300 yards apart and at a speed of 30 knots when they suddenly, and without any change of engine speed or helm on the part of either ship, found themselves literally side by side, a matter of feet apart. The two captains managed to separate their ships without an actual collision but it was a close-run thing for which there was no contemporary explanation. In the second incident the cruiser HMS *Curacao* was 'sucked' under the bow of the passenger line RMS *Queen Mary*, sliced in two and sunk with considerable loss of life. Then, in the 1960s and early 1970s, a series of unexplained accidents occurred in the United States in which tugs working with the then new type of Very Large Container Carriers were hit and then capsized by the larger ships for no apparent reason.

The matter had been studied in academic institutes, and as early as 1949, in a paper presented to the Royal Institute of Naval Architects, Professor A. M. Robb had shown a diagram in which 'the large ship overlaps the after end of the small ship, and two regions of increased pressure coincide. In such circumstances it is to be expected that the small ship will be deflected across the bow of the larger ship.'[8]

Admiralty documents had also identified the problem. BR 1742/52, issued in 1952, stated that: 'the best position for ships close aboard is exactly abeam or, if the receiving ship is very much smaller than the supplying ship, in the intermediate area between the bow and stern pressure zones and at least 80 feet from the supplying ship, where the interaction effects on ship handling are small'.

In other words, even though in 1976 far less was known of 'interaction' than is the case today, the danger of two ships, one significantly smaller than the other, steaming close together was appreciated, even if it was not properly understood. Furthermore, the Royal Navy's own pamphlet on the subject advised that such ships should maintain a distance of at least 80 feet.

However, in the manoeuvre conducted on 20 September 1976, the minesweepers were instructed to use heaving lines, and even the Admiralty's own Manual of Seamanship stated that experienced sailors would rarely throw 70 feet. *Fittleton's* sailors were reservists, not regulars. So, though capable and well-trained, they could scarcely be expected to be as skilled. In fact, as was proved on the day, they found 40 feet impossible.

Owing to her original design requirements, *Mermaid's* transfer point was considerably further forward than on other frigates. Indeed, its location so far forward took several of the other minesweeper commanding officers by surprise, for they had expected it to be approximately amidships. After the first collision Paget managed to stabilize *Fittleton* briefly, but having done so his ship started to surge forward – he told the court martial that he 'had the sensation we were surfing' – which was probably because she was riding atop the diverging wake (that is, the bow wave) from *Mermaid's* bow.

It was at this point, according to his evidence, that Paget realized that his ship was still moving forward relative to the frigate, and nothing he did seemed to slow her down. Since doing nothing was not an option, he decided that the only alternative was to try to speed up and turn to port, and he issued the necessary orders. With hindsight, such an increase in speed might possibly have served to exacerbate the problem, but since the rudders had no effect it seems possible that the propellers were having no effect either. Thus, matters had already passed the point where anything done by Paget could have had any effect over what happened next.

Fittleton, surging forwards parallel to *Mermaid* but at a higher speed, eventually reached a position where her stern quarter was alongside *Mermaid's* bow quarter. The high pressure in these regions produced a force that pushed *Fittleton's* stern to port (that is to say, away from *Mermaid*) but that had little effect on the bow of the much larger frigate. This turned the smaller minesweeper towards and under the much larger frigate's bow. It is also significant that *Fittleton* did not start to heel until she was hit by *Mermaid's* bow. This fact, as stated by Paget in his evidence and not challenged by the prosecution, is a strong indication that the minesweeper was not under helm at the time. All these events were the consequence of *Fittleton* being so close to *Mermaid* and, despite his acknowledged skill and experience, were well beyond Paget's control.

It must also be borne in mind that, although the ships were doing only 11 knots, things were now happening very fast aboard *Fittleton*. Because of the relative sizes of the two ships, *Fittleton*'s response to the force would have occurred in about seven-tenths of the time for a similar response from *Mermaid*. The magnitude of *Fittleton*'s response would have been about six times greater than that of *Mermaid*. In other words, both the perceived pace and the magnitude of events were much greater in the smaller ship.[9] Thus, Paget's very honest confession that he had a 'blind spot due to disorientation' was in fact the literal truth.

In the light of the hydrodynamic problems that occur when two ships of disparate size operate very close together, problems which, as has been explained, were known at the time, three circumstances might have prevented tragedy. First, a line-throwing gun, a device which has a much greater range than a human thrower, might have been used, thereby enabling the minesweepers to stand much further away from *Mermaid*. All minesweepers carried such a gun, but it would appear that its use was never considered.

Second, Paget could have declined to undertake the manoeuvre. At the court martial the admiral suggested that if Paget 'was unhappy about the situation he could have hauled down his signal indicating that he was not ready' and that 'if any of the RNR commanders were unhappy about an exercise because of inexperience I would have expected them to say so'. This is a somewhat ingenuous interpretation of the situation. The RNR captains were on their annual training exercise and it was only human nature that they should be on their mettle and feel the need to prove themselves to their crews and, in particular, to their admiral. Certainly, a captain is responsible for his ship, but when an order is given by a superior officer it is reasonable to assume that the senior officer knows what he is doing and has some perception of the consequences of his actions. Further, Paget was in command of the leading ship.

Had he, for whatever reason, declined to comply with the order it would have been a very public act that would, in addition, have caused utter confusion among the other minesweepers. What might have happened subsequently can only be a matter for speculation.

In addition, two of the earlier signals had been peremptory in tone. First, there had been the rapid execution of the signal ordering a change in formation on *Mermaid*'s arrival, which led to some milling around as the minesweepers changed not only formation but also the order in which they were sailing. Then there was the 'What are you waiting for?' which not only indicated a degree of impatience but also caused further confusion by being incorrectly coded, since the three leading ships all thought that it was intended for them.

Third, given that the difficulties of the manoeuvre – the problem of interaction between two ships of disparate sizes, the unusually forward location of the transfer position aboard *Mermaid* and the actual (as opposed to theoretical) throwing capability with a heaving-line – were either known or should have been known, the manoeuvre should not have been attempted in the first place.

The Admiralty letter quashed the charge on which Paget was found guilty. However, the letter also stated that the Board had 'concluded that while Lt-Cdr Paget hazarded *Fittleton* due to errors of judgement based on lack of experience, these did not amount to negligence under section 19 of the Naval Discipline Act'.

Paget was certainly not 'inexperienced'. A civil Master Mariner, the highest grade in that profession, he had also served for sixteen years in the RNR and had considerable experience in handling a Ton-class minesweeper. His skill as a shiphandler was also widely acknowledged. It might have been that the Admiralty Board was attempting to suggest that Paget was inexperienced in the particular manoeuvre being attempted, but if that was the case, the letter did not make it clear. It is also difficult to discover exactly what the Board was referring to when it used the phrase 'error of judgement', since a detailed examination of the circumstances indicate that, given the situation in which he was placed, Paget did everything that was within his power to do.

The responsibilities of a reservist commanding officer when called to duty are particularly demanding, because as soon as he (or she) dons a uniform, the fact that he is a civilian only temporarily on military duty becomes an irrelevance. Whether captaining a ship, commanding an army unit, or leading an airforce squadron, the reservist's responsibilities are no different from those of a professional in the same field, as would be the case in war.

Whether the ship is an 80,000-ton aircraft-carrier or a 300-ton minesweeper, or whether the unit is composed of 1,000 men or 300, the commanding officer's responsibilities are very precise. He or she bears the full responsibility for the unit's efficiency, for the proper and timely performance of any task placed upon it, and for the safety of both the unit and its members. When the unit is placed in a situation of potential hazard, those responsibilities become, if anything, even greater, with everything depending upon the professional skill, judgement and technical expertise of the commanding officer. However, an equally rigorous responsibility rests with the higher command to ensure that the demands placed upon commanding officers – whether regulars or reservists – are both practically and technically feasible.

13

Went the Day Well?

The commanding officer's position is *sui generis* an appointment without parallel, for its holder exercises greater direct power over the men and women in that unit than any other individual in the military chain of command. It is, too, a position that normally attracts considerable prestige, as well as some benefits and privileges, but, as has been shown in the preceding chapters, these pale into insignificance when compared with the obligations and responsibilities that it entails. It can also be exercised at a remarkably young age. Lieutenant-Colonel Tillet, for example, was considered to be a 'very experienced officer' when he took command of the 2/Devons in July 1917 at the age of 24, while Wing-Commander Gibson took over at 106 Squadron, his first independent command, at 23 and just one year later was entrusted with raising a new unit – 617 Squadron – from scratch for an exceedingly ambitious operation, and against a very tight timetable.

While it is true that some fighting forces are driven by a political belief, most are motivated by traditional factors such as patriotism, a desire to defend the homeland or hatred of the invader, or, at the most basic level, because they have been conscripted and have little choice in the matter. Whatever the personal motivation, it is the commanding officer's task to mould these disparate personalities by inspiring them with a common purpose and thus weld them into a cohesive whole.

In times of peace the commanding officer is wholly responsible for the safety of his unit and the well-being of his troops, and for the preparation of the unit for its combat role, although, as Lieutenant-Colonels Austin and Stewart learnt at Koh Tang and in Bosnia, this may not take the form that might reasonably have been expected. Once engaged in any form of conflict, however, the commanding officer assumes even greater responsibilities, deploying the unit to achieve the mission it has been given by higher command, and in doing so bearing almost total responsibility for the lives of all subordinates in the unit.

The decisions which have to be made are frequently hard and sometimes appalling. Dealing out death and destruction to the enemy is one thing, but knowingly to cause the death of people on one's own side is quite a different matter. Commander Egan, for example, was faced with a dreadful dilemma when it became clear that the Japanese submarine was beneath survivors of the *Khedive Ismail* struggling in the water. Perhaps even worse is to have to make decisions which must inevitably cause the death of members of one's own unit, as happened when Lieutenant-Commander Esmonde and his fellow airmen approached the German battlecruisers, or when Lieutenant-Colonel Anderson-Morshead of the 2/Devons and his men faced the German attack. Each bore an awesome responsibility, but neither shirked it, and their men followed them unflinchingly.

There is no standard mould for commanding officers. They must be placed according to their capabilities and personalities, and a man or woman who might be entirely suitable for one command appointment might be much less successful, even disastrous, in another. This is, however, a matter of selection, based on detailed study of previous performance and experience, although even then mistakes can be made.

In most armed forces it is obligatory for an officer to fill a commanding officer's appointment before progressing to command at higher levels, and there are sound military reasons for this. Experience shows, however, that a thoroughly good commanding officer can sometimes make an indifferent senior officer, while many effective admirals and generals have not necessarily been particularly successful commanding officers. During the 1990s, for example, many British newspapers carried obituaries of Second World War officers who were described as having been outstanding commanding officers of units such as destroyers, battalions and airforce squadrons. They had stayed in the Services after the war, only to leave in the 1950s or early 1960s having subsequently achieved one or at most two further steps in promotion. Clearly, the factors that had made them such good fighting commanders had not necessarily made them so suitable for the different battles that have to be fought in peacetime.

The commanding officer who is most frequently overlooked by the history books is the one who keeps the machine going. An undoubted hero such as Lieutenant-Colonel Anderson-Morshead of the 2/Devons is deservedly remembered, but Sunderland and Tillett, who died 'going over the top' with the same battalion, are equally deserving of recognition. Then, too, the fame of Wing-Commander Gibson of 617 Squadron is assured, but Wing-Commanders Littler and Goodman, who died leading 103 Squadron into battle, and the other commanding officers who sustained the unit through some very bleak times were equally heroic, albeit in a different way. Nor are competence and endurance the prerogatives of one side in war, as the careers of the U-boat commanders Kapitän-leutnants von Forstner and Lüdden show. It is commanding officers such as these who keep the military wheels turning, who provide the inspiration which keeps their troops and crews going and who lead them into battle time after time.

Many of the acute problems that arise to confront commanding officers are not the result of some cataclysmic event or an enormous and totally unexpected blunder, but are, rather, the outcome of a series of minor errors or ill-judged decisions which lead slowly but inexorably to disaster – a long and gentle but nevertheless slippery slope. Some of those decisions may have been taken many years before. That the Royal

Navy's P-class destroyers should mount 4-inch rather than the more usual 4.7-inch guns must have seemed a rational or simply a more economical decision at the time it was made by planners in London. Nevertheless, as a result of that decision two destroyers would, four years later, be unable to draw the correct ammunition from the depot in Ceylon, with the result that a few days later they were unable to use their guns to sink a Japanese submarine.

The deaths of Lieutenant-Commander Esmonde and so many men of 825 Squadron can be seen in the same light. The higher command gradually whittled away the forces intended to attack the German battlecruisers should they attempt to withdraw up the Channel to Germany. As a consequence, when the ships did finally make the attempt, they were faced by only a few torpedo-boats, six obsolete Swordfish and a host of ineffective and unrealizable plans. It is not, however, the planners and the staff officers who carry the consequences of such decisions, but the commanding officers and their units.

Other ill-judged decisions are taken closer to the action. The repeated failure to order that convoy KR8 should zigzag, for example, was almost certainly a major contributory factor in the loss of the *Khedive Ismail*. Each time the subject arose there was a seemingly valid reason for postponing the decision, until it was simply too late.

Special forces units raise a particular problem. Their operations are bound, by definition, to be daring and unconventional, and if they succeed all is well and good. There have been some spectacularly successful hostage rescues by national special forces – by the Germans at Mogadishu, Somaliland, in October 1977, by the Israelis at Entebbe, Uganda, in July 1976, and by the United Kingdom in the Iranian Embassy siege in 1980. These were all daring missions, and on their completion the units involved and their commanding officers were praised accordingly. Things can, however, go badly wrong as was the case with the attempted rescue of the American hostages in the Tehran Embassy in 1980 and, as described in Chapter 6, with the Italian attack on Malta in 1941. In the latter case, though there is no doubt that the Italian Navy's Decima unit was both courageous and determined, the plan was fatally flawed from the start. Bravery is not sufficient to overcome a bad plan.

There can be very few commanding officers, whether on land or sea or in the air, who did not feel a personal involvement with their units. Langsdorff summed up the attitude of the great majority when he wrote, in his farewell letter, that 'for a captain with a sense of honour, it goes without saying that his personal fate cannot be separated from that of his ship'. However, while virtually all commanding officers would share that sentiment, few would go so far as to take their own lives to demonstrate the fact. The practice of commanding officers committing suicide or, in the naval case, of going down with their ships does not seem to possess any particular merit.

The subject of war crimes is contentious and deserves fuller study than can be given here, but it is one of special concern to commanding officers. There can be no doubt that Kapitänleutnants Eck and Patzig behaved in a vicious and cruel manner when they and their crew spent many hours hunting down the few remaining survivors of the *Peleus* and the *Llandovery Castle*, while Captain Ariizumi's behaviour in the Indian Ocean was even worse, showing an inexcusable and in many ways an inexplicable cruelty. There is, however, an irony in Eck's case. If the entire crew of the *Peleus* had died as an immediate result of the torpedo strike or if he had simply sailed away and abandoned them to their fate, he would have behaved no better and no worse than the majority of submarine commanders in all combatant navies in both world wars.

The modern emphasis on war crimes is something that puts commanding officers of all Services on their guard. Following the Falklands War, for example, a small number of people conducted a long campaign in which they sought to demonstrate that the Argentine cruiser *General Belgrano* had been sunk by the British nuclear-powered attack submarine HMS *Conqueror* outside the exclusion zone and that it was therefore not a legitimate target. Had they proved their case, the submarine's commanding officer might have been liable for trial as a war criminal. Similarly, if it had been proved that the commanding officer of USS *Nicholas* had fired on a unit trying to conduct a collective and formal surrender, then he too might have been charged with a war crime.

The case of Captain Mayazumi of the Japanese cruiser *Tone* raises the question of just how far a serviceman is bound to obey his superior's orders – a consideration of great concern to commanding officers, both in their obedience to their superiors and in their subordinates' acceptance of orders they give. Mayazumi raised a string of objections to the orders he received from his admiral, and both he and his executive officer did everything that was reasonably within their power to circumvent the admiral's orders. But when Mayazumi issued his own orders to the master-at-arms to execute the prisoners in his charge, he would brook no disobedience.

Command is a lonely experience and even the most self-sufficient commanding officer may have his private moments of doubt and fear. Eugene Esmonde, though supported by his Christian faith, was described as 'ashen-faced' as he walked towards his aircraft on what proved to be his last mission. And even Guy Gibson, outwardly the most self-confident and ebullient of leaders, betrayed his feelings on one occasion. Many years after the war Mary North, an RAF nurse who had become close to Gibson, described how one night, following a particularly difficult flying operation over Berlin and the loss of two friends, Gibson arrived at her hospital. Like Esmonde, he looked ashen. As they sat in his car Gibson started to shake silently but uncontrollably. Mary North took him in her arms, but even so it was thirty minutes before she was able to calm him down.[1] How many other apparently imperturbable leaders have known equal despair in the long, dark stretches of the night?

Captains at sea, for whom there is no respite, face even greater loneliness. Kapitän zur See Langsdorff was at sea for five months aboard *Graf Spee*, while the U-boat commander Kapitänleutnant Lüdden endured nearly a year in command without a break. The loneliness of command, however, cannot be measured solely in terms of time. Though Langsdorff was able to exchange signals with Germany and to talk to the local German ambassador while his ship lay in Montevideo harbour, he alone bore the responsibility for the dreadful decision about the future of his ship. Langsdorff had several days to ponder that decision, but Lieutenant-Commander Paget had only a very few seconds to decide what to do as *Fittleton* rushed along *Mermaid*'s side.

There are many thoroughly capable commanding officers, but there are also a gifted few who possess that 'certain touch' – a genius for war – that sets them apart from the rest. Such people are capable not only of recognizing the fleeting opportunity but also of deploying their troops to take advantage of it, thereby raising command from a skill to an art. Despite their many successes elsewhere, Colborne at Waterloo, Gibson over the Ruhr dams, and Lownds at Khé Sanh never knew a finer moment – it was as if everything that had gone before was simply preparation, and all that followed was mere anti-climax.

There are, of course, wicked, misguided, dishonest or ineffective commanding officers who somehow elude the safeguards built into their forces' selection systems, but they are a minority, as in any segment of society. The vast majority of commanding officers are honest and reasonable men who try their very best in what is the most demanding and responsible appointment in the armed forces. Whatever side they are on, they fight as well as they can for a cause in which they believe, and though they may have different interpretations of the meaning of 'freedom', they would probably seek no better epitaph than this:

> *Went the day well?*
> *We died and never knew,*
> *But, well or ill,*
> *Freedom, we died for you.*

Notes

Chapter 1: A Many-talented Officer

1. John Sweetman, *The Dambusters' Raid*, (Arms & Armour Press, 1990), p. 70.

Chapter 2: 'A Genius for War'

1. 'An Unpublished Memoir', Regimental Chronicle of the 52nd Foot, pp. 111–33; G. C. Moore Smith, *Life of John Colborne, Field Marshal Lord Seaton*, (John Murray, London, 1903); and Lieutenant Frederick Hope Patterson, *Personal Recollections of the Waterloo Campaign*, (reprint, Waterloo Committee, 1997).
2. 'This Month's Cover', *US Naval War College Review*, Vol. XLVI, No. 2, Spring 1993, pp. 116–18; and PRO ADM 52/1999 (Master's Log, HMS *Solebay*, September 1776).
3. Lieutenant-General Willard Pearson, 'The War in the Northern Provinces, 1966–1968', *Vietnam Studies* (Department of the Army, Washington, DC, 1975); and Bernard C. Nalty, *Air Power and the Fight for Khé Sanh*, (Office of Air Force History, Washington, DC, 1973).
4. John Sweetman, *The Dambusters' Raid*, (Arms & Armour Press, 1990).

Chapter 3: 'Warriors for the Working-Day'

1. W.J.P. Aggett, *The Bloody Eleventh*, Vol. III, 1914–1969, (The Devonshire and Dorset Regiment, Exeter, 1995); and C.T.

247

Atkinson, *The Devonshire Regiment 1914–1918*, (Eland Bros., Exeter, 1926).

2. The Type XIV was specially designed to carry diesel fuel, ammunition and supplies for operational U-boats, the transfers being carried out on the surface.

3. *U487* was sunk on 13 July 1943 by aircraft from the carrier USS *Core*.

4. U-boats had constant trouble with their torpedoes. In this case, the tropical heat caused a deterioration in batteries, resulting in slow running.

5. It comprised: tin (109,877kg): rubber (12,856kg), hides (812kg), quinine (500kg), opium (200kg), and wolfram (18,616kg).

6. *U1062*, a Type VIIF, was on its way to deliver thirty-nine torpedoes to the base at Penang.

7. *U66* was desperately short of fuel and its crew exhausted. At the time of *U188*'s approach *U66* was being hounded by the US carrier *Block Island*. She was eventually sunk by the destroyer-escort *Buckley* on 6 May.

8. Lüdden subsequently served in various staff appointments but was killed in a fire aboard the accommodation ship *Daressalam* at Kiel on 13 January 1945. Material in this chapter has been drawn from R. Busch and H.-J. Röll, *German U-Boat Commanders of World War II* (Greenhill Books, London, 1999); and K. Wynn, *U-Boat Operations of the Second World War*, Vol. 1; (Conway Maritime Press, London 1997).

9. R. Busch and H.-J. Röll, *German U-Boat Commanders of World War II*, (Greenhill Books, London, 1999); K. Wynn, *U-Boat Operations of the Second World War*, Vol. 1; (Conway Maritime Press, London, 1997); and personal research in the Royal Naval Submarine Museum, Gosport, and in the *U-Boat Archive*, Altenbruich, Germany.

10. Sid Finn, *Black Swan* (Newton Publishers, Edenbridge, 1989); R. Overy, *Bomber Command, 1939–1945* (HarperCollins, London, 1997); P. Moyes, *Bomber Squadrons of the RAF and their Aircraft* (Macdonald, London, 1964); and Mrs G. Westrup, former Chairman, RAF Elsham Wolds Association.

11. 'Let me speak proudly: tell the Constable we are but warriors for the working-day.', *Henry V*, Act 4, scene 3.

Chapter 4: Two Fateful Decisions

1. In nautical terms, speed of advance is the rate at which a ship is approaching its destination, whereas the ship's speed is the rate at which it is moving through the water. Thus, when zigzagging, a ship might be moving at a speed of 13 knots but its rate of advance would be somewhat slower, probably about 11 to 11½ knots.
2. These were in Admiralty CB.04024, Art.33a, and CB.03058, Art.34, respectively.
3. Letter marked 'Top Secret & Personal' to Captain Josselyn, and signed H. V. Markham, Secretary to the Board of Admiralty, quoted in B.J. Crabbe, *Passage to Destiny* (Paul Watkins, Stamford, 1997), p. 100.
4. It was believed by some former members of *Petard*'s crew that the scene in Nicholas Monsarrat's novel *The Cruel Sea*, in which a corvette attacks through swimming survivors, was based upon this incident in *Petard*'s fight against *I27*.
5. Personal research in the PRO including files ADM 53/119538, ADM 199/526, ADM 199/2052, ADM 199/2133 and ADM 199/2147; B.J. Crabbe, *Passage to Destiny* (Paul Watkins, Stamford, 1997); extract from Captain Bailey's proposed autobiography in Lt.-Cdr. L.G. Hooke, *Ooops: Autobiography of Gordon Hooke* (J.&K.H. Publishing, Hailsham, 1997), Part 2, pp. 218–22); *Royal Artillery Commemoration Book, 1939–1945* (Royal Artillery Institute, Woolwich), pp. 346–50; G.G. Connel, *Fighting Destroyer* (William Kimber, London, 1976); J. Rohwer, *Axis Submarine Successes, 1939–45* (US Naval Institute Press, Annapolis, 1983).

Chapter 5: To the Last Man and the Last Round

1. To be strictly accurate, the RAF provided the aircraft, the pilots and most of the mechanics; the Royal Navy provided the observers and radio operators.
2. Captain E. Brown, RN, *Wings of the Navy* (Jane's Publishing Company, 1980), p. 19.
3. The crucial damage to the *Bismarck* was inflicted by a subsequent attack by a different squadron.
4. Chaz Bowyer, *Eugene Esmonde, VC, DSO* (William Kimber, 1983), p. 111.

5. Bowyer, p. 168.
6. Chaz Bowyer, *Eugene Esmonde, VC, DSO* (William Kimber, 1983); and Captain E. Brown, RN, *Wings of the Navy* (Jane's Publishing Company, 1980).
7. W.J.P. Aggett, *The Bloody Eleventh*, Vol. III, 1914–1969 (The Devonshire and Dorset Regiment, Exeter, 1995); and C.T. Atkinson, *The Devonshire Regiment, 1914–1918* (Eland Bros., Exeter, 1926.
8. Andrew Gordon, *The Rules of the Game: Jutland and the British Naval Command* (John Murray, 1996); and H.W. Fawcett and G.W.W. Cooper, *The Fighting at Jutland* (n.p., 1920).

Chapter 6: Bravery is not Enough

1. Long-range radar had, in fact, spotted a 'blip' off the coast of Sicily. This was *Diana*, but when it stopped and then returned towards Sicily it was interpreted as not being a threat. None of the smaller craft showed up on the primitive radar. Ultra had suggested that an attack was planned against Malta, but there was no indication of the form it would take.
2. On 19 December 1941 three human torpedoes were launched from the submarine *Scirè*, commanded by Prince Borghese. Entering the British naval base at Alexandria, they caused very severe damage to the battleships *Valiant* and *Queen Elizabeth*, both of which were out of action for many months to come, and to a destroyer and a tanker.
3. Excerpts from an article by Admiral (Retired) Antonio Scialdone, a former member of Decima, quoted in a letter from Dr Ezio Bonsignore to the author.
4. PRO ADM 223/460, ADML 413/199, WO 169/3251, WO 169/3280; Prince J. Valerio Borghese, *The Sea Devils* trans. James Cleugh (Andrew Melrose, 1952); Richard O'Neill, *Suicide Squads* (Salamander Books, 1981), pp.77–83; and *All the World's Fighting Ships: 1922–1946* (Conway Maritime Press, 1980), pp. 312–16.

Chapter 7: War Crimes

1. *The Times*, 16 July 1921, p. 9ff.
2. The Times, 14, 16 and 18 July 1921; John Cameron (ed.), *The Trial*

of Heinz Eck, August Hoffmann, Walter Weisspfennig, Hans Richard Lenz and Wolfgang Schwender (The Peleus *Trial)*, (William Hodge, 1948), Appendix IX, pp. 167–83; and British Government White Paper, Comd. 1450/21.

3. The records of the *Peleus* case are held in the Public Record Office, London, in PRO/WO311/219, PRO/WO/235/604 and PRO/WO/233/5, the last file including the complete short-hand transcript of the trial. The trial is also recorded in John Cameron (ed.), *The Trial of Heinz Eck, August Hoffmann, Walter Weisspfennig, Hans Richard Lenz and Wolfgang Schwender (The* Peleus *Trial)*, (William Hodge, 1948).

4. There are two possible explanations for these men not going below. First, inside the hull were some one hundred very angry and disappointed Japanese, and the US sailors would have been heavily outnumbered and vulnerable to physical assault, maybe even death. The second is that on a long patrol the inside of a submarine inevitably develops a strong atmosphere which may not be noticeable to the crew but is very offensive to outsiders, and cleanliness of his boat was not one of Ariizumi's priorities.

5. Imperial War Museum Boxes BMWC 1–7 (documents assembled by the British Minor War Crimes Commission in Japan, 1945–8); and PRO files ADM 199/2147, 199/526 and 199/255.

6. The account of this episode is based on primary sources in the Public Record Office, Kew, including: the full proceedings; the transcript of the shorthand record of the trial (PRO/WO 235/1089); various survivors' statements and correspondence between GHQ FARELF and the War Office, London, in 1949 (PRO/WO 311/549); and other survivors' statements, including a selection of photographs (PRO/WO 311/564).

Chapter 8: Surrender and Receiving Surrender

1. Barrie Pitt, *The Crucible of War: Western Desert, 1941* (Jonathan Cape, 1980), pp. 187–8.

2. Captain J.D. Boswell, *The Capture of the Diamond Rock, 1803, by an Eyewitness* (no publisher, 1833); PRO ADM 1/325; and V. Stuart and G.T. Eggleston, *His Majesty's Sloop of War Diamond Rock* (Robert Hale, 1978).

3. John Ashby, *Seek Glory, Now Keep Glory* (Helion, Solihull, 2000).
4. John Ashby, *Seek Glory, Now Keep Glory* (Helion, Solihull, 2000).
5. The proceedings of the subsequent board of inquiry were released to the author on the condition that the captain should not be identified.
6. 'Report of the Fact-finding Body Required to Conduct a Hearing into the Facts and Circumstances Surrounding the Attack on Iraqi Oil Platforms by USS *Nicholas* (FFG 47) While Operating in the Northern Arabian Gulf on or About 18 January 1991, CinC US Atlantic Fleet 5830, dated 16 August 1991; and Rear-Admiral Horace B. Robertson, Jr., 'The Obligation to Accept Surrender', *US Naval War College Review*, Spring 1993, pp. 102–15.

Chapter 9: Suicide

1. R.N. Buckley (ed.), *The Napoleonic War Journal of Captain Thomas Browne: 1807–1816* (Bodley Head/Army Records Society, 1987), pp.142–3.
2. Following a German inquiry, a number of French dockyard workers were later executed.
3. *U505* undertook two further cruises, the second of which culminated in her capture at sea by a US Navy task group, following which she was towed to the USA. She is now on permanent display in Chicago.
4. Personal research in the relevant files at the U-Boat Archive, Altenbruich, Germany. K. Wynn, *U-Boat Operations of the Second World War* (Conway Maritime Press, 1997) Vol. 1 and 2; and C. Blair, *Hitler's U-Boat War* (Weidenfeld & Nicolson, 1997), Vols. 1 and 2.
5. After the war, *Sculpin*'s survivors related what he had done, but without knowing the reason why. The Pentagon, however, did and he was posthumously awarded the Congressional Medal of Honor.
6. It was popularly believed at the time that, before committing suicide, Langsdorff had wrapped himself in the battle ensign of the former Imperial German Navy rather than that of the contemporary Kriegsmarine of the Third Reich. This appears to have been an example of wartime 'disinformation' and there is no evidence whatsoever to support such a claim, for not one eyewitness

account by those who discovered his body suggests that he used the 'old' rather than the 'new' flag.
7. PRO ADM 1/9759.
8. PRO ADM 223/69, ADM 1/9759, ADM 223/68, ADM 186/794 and ADM 116/4180; R.E. Laurence, *Desde Wilhelmshaven a Montevideo* (Rosario, Argentina, 1992); and Sir Eugen Millington-Drake, *The Drama of* Graf Spee *and the Battle of the Plate* (Peter Davies, 1964).

Chapter 10: The Ultimate Demonstration of Responsibility

1. Edward Giffard, *Deeds of Naval Daring* (John Murray, 1884), pp. 243–4.
2. Ten men were rescued but only a handful of these survived the rigours of a three-year stay in a Japanese prisoner-of-war camp, so that news of the engagement did not reach England until late 1945. In late 1946 Wilkinson was awarded a posthumous Victoria Cross.
3. *The Queen's Regulations and Admiralty Instructions for the Government of Her Majesty's Naval Service* (Her Majesty's Stationery Office, 1862), p. 333.
4. *Regulations for the Government of the Navy of the United States* (Government Printing Office, Washington, DC, 1905), Regulations 423 (1) and (2). The current document, *United States Navy Regulations, 1990* (Department of the Navy, Washington, DC, Regulations 0852.1 and 0852.2, is essentially identical.
5. Andrew Gordon, *The Rules of the Game* (John Murray, 1996), p. 258.
6. Sir Eugen Millington-Drake, *The Drama of* Graf Spee *and the Battle of the River Plate*, p. 366 (Peter Davis, 1964), quoting from Dudley Pope, *The Battle of the River Plate* (William Kimber, 1956).
7. Hashimoto, commanding officer of *I8*, was flown to the United States to appear as the main prosecution witness at McVay's court martial.
8. It has been suggested by some sources that Captain I. North, Master of the container ship *Atlantic Conveyor*, went down with his ship when it was hit by two Exocet missiles on 25 May 1982 during the Falklands War. As with a number of similar instances, this can neither be proved nor disproved.

Chapter 11: A Sort of Peace

1. 'Fourteen Hours at Koh Tang', *USAF South-East Asia Monographs*, Vol. 3, No. 5, ed. Major A.J.C. Lavalle (US Government Printing Office, Washington, DC, September 1975); 'Individual heroism overcame awkward command relationships, confusion and bad information off the Cambodian coast', *Marine Corps Gazette*, October 1977, pp. 24–34; and 'The Assault on Koh Tang', *Navy Times*, 27 August 1975, pp. 15 and 33.
2. In British military parlance a 'battalion group' is based on an infantry battalion, but with additional units under command for a particular task. Such additions will normally be a squadron of cavalry, an engineer troop and detachments from various logistical units.
3. Lieutenant-Colonel Bob Stewart, *Broken Lives* (HarperCollins, 1993); and personal conversations and correspondence with Colonel Stewart.
4. *Broken Lives* (HarperCollins, 1993), p. 317.

Chapter 12: The Loss of HMS Fittleton, *20 September 1976*

1. The official documents concerning this disaster will not be released until 2006. This chapter is based upon: contemporary newspaper reports, including those in *The Times* and the *Daily Telegraph*; a considerable number of interviews and correspondence with survivors and with eyewitnesses from a number of ships at the scene of the disaster; study of many professional papers held in the Public Record Office and by bodies such as the Royal Institute of Naval Architects; and discussions with experts in hydrodynamics and ship behaviour.
2. To be strictly accurate, the Ton-class ships were officially described as Mine Counter-Measure Vessels (MCMV), a term which embraced both minesweepers and minehunters. Since all the ships referred to in this chapter were minesweepers, this rather less inelegant term will be used instead.
3. The number of RNR Ton-class ships was reduced to eleven when Humber Division closed in 1958. When an RNR ship needed a refit, it was replaced by another vessel, thus ensuring that the RNR always had the necessary number of hulls with which to train.

4. These conditions were described in the 'Circumstantial Letter' presented to the Court Martial by the Prosecuting Officer' (*Daily Telegraph*, 22 February 1977).
5. Statement presented to the court martial in the prosecution's 'Circumstantial Letter' (*Daily Telegraph*), 22 February 1977).
6. *Admiralty Manual of Seamanship*, 1964, Vol. 1, p. 124.
7. *Daily Telegraph*, 12 May 1977.
8. Professor A.M. Robb, Professor of Naval Architecture, Glasgow University, paper on 'Interaction between Ships' presented at the Institute of Naval Architects in 1949 (*Transactions of the Institute of Naval Architects*, 1949, Vol. 9, No. 4, p. 338.
9. Information supplied by Dr D. Fryer.

Chapter 13: Went the Day Well?

1. Richard Morris, *Guy Gibson* (Viking, 1994), p. 132.

Bibliography

Books

All books were published in London unless otherwise stated.

Aggett, W.J.P., *The Bloody Eleventh* (The Devonshire and Dorset Regiment, Exeter, 1995), Vol. III, 1914–69

All the World's Fighting Ships: 1922–1946 (Conway Maritime Press, 1980), pp. 312–16

Ashby, John, *Seek Glory, Now Keep Glory* (Helion and Co., Solihull, 2000)

Atkinson, C.T., *The Devonshire Regiment 1914–1918* (Eland, Exeter, 1926)

Blair, C., *Hitler's U-Boat War* (Weidenfeld & Nicolson, 1997), Vols. 1 and 2

Borghese, Prince I. Valerio, *Decima Flottiglia MAS*, trans. James Cleugh, *The Sea Devils* (Andrew Melrose, 1952)

Boswell, Captain J.D., *The Capture of the Diamond Rock, 1803, by an Eyewitness* (n.p., 1833)

Bowyer, Chaz, *Eugene Esmonde, VC, DSO* (William Kimber, 1983)

Brown, Captain E., *Wings of the Navy* (Jane's Publishing Company, 1980)

Buckley, R.N. (ed.), *The Napoleonic War Journal of Captain Thomas Browne: 1807–1816* (Bodley Head/Army Records Society, 1987)

Busch, R., and Röll, H.-J., *German U-Boat Commanders of World War II* (Greenhill Books, 1999).

Cameron, John (ed.), *The Trial of Heinz Eck, August Hoffmann, Walter Weisspfennig, Hans Richard Lenz and Wolfgang Schwender (The Peleus Trial)* (William Hodge, 1948)

Connel, G.G., *Fighting Destroyer* (William Kimber, 1976)

Crabbe, Brian J., *Passage to Destiny* (Paul Watkins, Stamford, 1997)

Dove, Captain Patrick, *I was* Graf Spee*'s Prisoner!* (Cherry Tree Books, 1940)

Fawcett, H.W., and Cooper, G.W.W., *The Fighting at Jutland* (n.p., 1920)

Finn, Sid, *Black Swan; The Story of 103 Squadron* (Newton, Edenbridge, 1989)

Giffard, Edward, *Deeds of Naval Daring* (John Murray, 1884)

Gordon, Andrew, *The Rules of the Game: Jutland and British Naval Command* (John Murray, 1996)

Hessler, Günter, *German Naval History: The U-Boat War in the Atlantic, 1939–1945* (Her Majesty's Stationery Office, 1989)

Hirschfeld, Wolfgang, *Hirschfeld, The Story of a U-boat NCO, 1940–1946* (Leo Cooper, 1996)

Hooke, Lt. Cdr. L.G., *Ooops: Autobiography of Gordon Hooke* (J&KH Publishing, Hailsham, 1997)

Laurence, R.E., *Desde Wilhelmshaven a Montevideo* (Rosario, Argentina, 1992)

Lavalle, Major A.J.C. (ed.), 'Fourteen Hours at Koh Tang', *USAF South-East Asia Monographs* (US Government Printing Office, Washington, DC, 1975), Vol. 3, No. 5

Millington-Drake, Sir Eugen, *The Drama of* Graf Spee *and the Battle of the River Plate, A Documentary Anthology* (Peter Davies, 1964)

Mitford, A.B., *Tales of Old Japan* (Macmillan, 1871), Vol. 1

Moore-Smith, G.C., *Life of John Colborne, Field Marshal Lord Seaton* (John Murray, 1903)

Morris, Richard, with Dobinson, Colin, *Guy Gibson* (Viking, 1994)

Moyes, P., *Bomber Squadrons of the RAF and their Aircraft* (Macdonald, 1964)

O'Neill, R., *Suicide Squads* (Salamander Books, 1981)

Patterson, Lieutenant Frederick Hope, *Personal Recollections of the Waterloo Campaign* (reprinted by the Waterloo Committee, 1997)

Overy, Richard, *Bomber Command, 1939–1945* (HarperCollins, 1997)

Pearson, Lt.-Gen. Willard, *Vietnam Studies, The War in the Northern Provinces, 1966–1968* (Department of the Army, Washington, DC, 1975)

Pitt, Barrie, *The Crucible of War: Western Desert 1941* (Jonathan Cape, 1980)

Röhwer, J., *Axis Submarine Successes 1939–1945* (US Naval Institute Press, Annapolis, 1983)

Röhwer, L.J., and Hummelchen, G., *Chronology of the War at Sea, 1939–1945* (Greenhill Books, 1992)
Royal Artillery Commemoration Book 1939–1945 (Royal Artillery Institute, Woolwich)
Stewart, Lieutenant-Colonel Bob, *Broken Lives* (HarperCollins, 1993)
Stuart, V., and Eggleston, G.T., *His Majesty's Sloop of War* Diamond Rock (Robert Hale, 1978)
Sweetman, John, *The Dambusters' Raid* (Arms & Armour Press, 1990)
Turner, E.S., *Gallant Gentlemen: A Portrait of the British Officer, 1600–1956* (Michael Joseph, 1956)
Wynn, Kenneth, *U-Boat Operations of the Second World War* (Conway Maritime Press, 1997), Vols. 1 and 2

Articles

'Individual heroism overcame awkward command relationships, confusion and bad information off the Cambodian coast', *Marine Corps Gazette*, October 1977, pp. 24–34
A.M. Robb, 'Interaction Between Ships', *Royal Institute of Naval Architects Transactions*, July 1949, p. 324–39
H.W.J. Chislett, 'Replenishment at Sea', *The Naval Architect*, July 1972, pp. 321–40
'The Assault on Koh Tang', *Navy Times*, 27 August 1975, pp. 15 and 33
'An Unpublished Memoir', *Regimental Chronicle of the 52nd Foot*, pp. 111–33
Rear-Admiral Horace B. Robertson, Jr., 'The Obligation to Accept Surrender', *US Naval War College Review*, Spring 1993, pp. 102–15
'This Month's Cover', *US Naval War College Review*, Spring 1993, pp. 116–118.

Official Documents

Admiralty Manual of Seamanship (Her Majesty's Stationery Office, 1952), Vol. II [B.R. 67(2/51)]
Admiralty Manual of Seamanship (Her Majesty's Stationery Office, 1964), Vol. III [B.R. 67(3)]

Bibliography

Admiralty documents: BR 1742/52, 'Replenishment Operations, Manoeuvring and Shiphandling,', 1952 (PRO ADM 234/393); CB.04024; CB.03058

British Government White Paper, Comd. 1450/21

Regulations for the Government of the Navy of the United States (Government Printing Office, Washington DC, 1905)

'Report of the Fact Finding Body Required to Conduct a Hearing into the Facts and Circumstances Surrounding the Attack on Iraqi Oil Platforms by USS *Nicholas* (FFG 47) While Operating in the Northern Arabian Gulf on or About 18 January 1991', Commander-in-Chief US Atlantic Fleet 5830, dated 16 August 1991

The Queen's Regulations and Admiralty Instructions for the Government of Her Majesty's Naval Service (Her Majesty's Stationery Office, 1862).

United States Navy Regulations, 1990 (Department of the Navy, Washington, DC, 1990)

Public Record Office, Kew

PRO ADM 1/325; ADM 1/9759; ADM 52/1999; ADM 53/119538; ADM 116/4180; ADM 186/794; ADM 199/255; ADM 199/526; ADM 199/2052; ADM 199/2133; ADM 199/2147; ADM 223/68; ADM 223/69; ADM 223/460; ADM 413/199; WO 169/3251; WO 169/3280, WO/233/5, WO/235/604; WO 235/1089; WO 311/219; WO 311/549; WO 311/564

Imperial War Museum

Boxes BMWC 1–7 (documents assembled by the British Minor War Crimes Commission in Japan, 1945–8)

Newspapers

The Times, 14, 16 and 18 July 1921, and various dates in 1976
Daily Telegraph, various dates in 1976

Index

Index